Digitization and Manufacturing Performance

Scrivener Publishing
100 Cummings Center, Suite 541J
Beverly, MA 01915-6106

Publishers at Scrivener
Martin Scrivener (martin@scrivenerpublishing.com)
Phillip Carmical (pcarmical@scrivenerpublishing.com)

Digitization and Manufacturing Performance

An Environmental Perspective

Edited by

Chandan Deep Singh

Department of Mechanical Engineering, Punjabi University, Patiala, Punjab, India

Talwinder Singh

Department of Mechanical Engineering, Punjabi University, Patiala, Punjab, India

and

Davinder Singh

Department of Mechanical Engineering, Punjabi University, Patiala, Punjab, India

Scrivener Publishing

WILEY

This edition first published 2025 by John Wiley & Sons, Inc., 111 River Street, Hoboken, NJ 07030, USA and Scrivener Publishing LLC, 100 Cummings Center, Suite 541J, Beverly, MA 01915, USA

For more information about Scrivener publications please visit www.scrivenerpublishing.com.

Wiley Global Headquarters
111 River Street, Hoboken, NJ 07030, USA

For details of our global editorial offices, customer services, and more information about Wiley products visit us at www.wiley.com.

Library of Congress Cataloging-in-Publication Data

ISBN 978-1-394-19776-7

Front cover images courtesy of Adobe Firefly
Cover design by Russell Richardson

Set in size of 11pt and Minion Pro by Manila Typesetting Company, Makati, Philippines

Printed in the USA

10 9 8 7 6 5 4 3 2 1

Contents

Preface

Currently, most production strategies are shaped by factors such as quality, cost, delivery, innovation, and responsiveness. Traditionally, firms have pursued these objectives by adopting production practices such as simultaneous engineering, increasing efficiency through defect elimination, setup reduction, and worker empowerment. Recent developments in the industry indicate another path to achieving production excellence: industry regulators and professional bodies should foster innovation across a broad spectrum of high-tech production facilities with environmental considerations in mind.

The industry's success hinges on the capabilities of production facilities and the competitive advantage gained through superior quality and reliability. This advantage translates into increased sales and developing a robust customer base, leading to greater market share and, ultimately, enhanced profitability, growth, and expansion. A firm's processes must exhibit competitive priorities that provide operational advantages to outperform competitors. These advantages are assessed, evaluated, and measured across parameters such as cost, quality, time, design, and flexibility. This book aims to meet the expectations of industrialists, policymakers, and academicians by evaluating the impact of production facilities, which are critical to the industry's success.

The manufacturing industry forms the material foundation and the backbone of a country's national economy, reflecting its comprehensive national strength. However, the growth in mechanical manufacturing has escalated the strain on global energy resources, producing wastewater, waste gas, and residues that significantly impact the environment. This is detrimental to the sustainable development of our ecological systems. Economic gains are no longer the sole objective of the manufacturing sector. Emphasizing ecological protection and the scientific utilization of resources now stand as critical goals.

In response, fostering the development of green manufacturing is crucial for the sustainable progress of the industry. The current manufacturing landscape still lacks adequate integration of intelligent technologies and sustainable green engineering practices, which are essential for maintaining a clean and superior production environment. This book introduces enabling mechanisms to assess the "greenness" of intelligent techniques employed in modern organizations. It also advocates for alternative renewable energy sources, zero-emission and zero-waste manufacturing facilities, alongside promoting green standards and awareness. It is anticipated that green engineering will emerge as a vital manufacturing standard that addresses the multifaceted issues of environmental, employee, and economic development globally.

Many industries still rely on conventional maintenance approaches, like breakdown maintenance, which do not align with sustainability performance indicators, thus operating at low efficiency. This book highlights the importance of Maintenance 4.0 techniques, which integrate sustainable key performance indicators (KPIs) in the maintenance decision-making process to optimize organizational and field service maintenance.

The need for this book arose from the recognition that green engineering, as a comprehensive approach, has not been sufficiently addressed. This study aims to explore and define the parameters contributing to the success of the manufacturing industry through green engineering initiatives. Adopting green engineering during the product development phase is crucial to minimize the industrial impact on the environment.

The book proposes different tools and techniques for sustainable product development, creating a new framework for Green Product Development (GPD) that can be readily implemented within organizations. In recent years, the automobile sector has taken significant steps toward reducing carbon emissions through new technologies that enhance fuel efficiency, explore alternative fuels, and develop zero-emission vehicles. For the automotive industry, sustainability is increasingly crucial to meet both regulatory and consumer demands. This book also explores various sustainable technologies specifically for the automobile sector, including gasoline-electric hybrid, electric, zero-emission hydrogen fuel cell, solar, and e-bio fuel cell vehicles. Green engineering has become an essential strategic tool for automobile companies to cultivate an environmentally conscious brand image and differentiate themselves from competitors in the market.

We want to express our deepest appreciation to all the contributors who have dedicated their time and efforts to making this book a success. Additionally, we extend our deepest thanks for the suggestions, help, and support from Martin Scrivener and the team at Scrivener Publishing.

The Editors
March 2025

1

Green Energy Technologies

Chandan Deep Singh

Punjabi University, Patiala, Punjab, India

Abstract
Green energy is that which comes from natural sources, such as the sun. Clean energy are those types which do not release pollutants into the air, and renewable energy comes from sources that are constantly being replenished, such as hydro power, wind power or solar energy. Clean energy technologies are renewable, less environmentally invasive ways of powering the global community. Some of the most common examples of clean energy sources include solar, wind, water, geothermal, bioenergy, natural gas, and nuclear power. The key with these energy resources are that they do not harm the environment through factor, such as releasing greenhouse gases into the atmosphere.

***Keywords*:** Green energy technologies, materials, biomass, electric power, hydro, solar, wind, geothermal

1.1 Introduction

As a source of energy, green energy often comes from renewable energy technologies such as solar energy, wind power, geothermal energy, biomass and hydroelectric power. Each of these technologies works in different ways, whether that is by taking power from the sun, as with solar panels, or using wind turbines or the flow of water to generate energy.

In order to be deemed green energy, a resource cannot produce pollution, such as is found with fossil fuels. This means that not all sources used by the renewable energy industry are green. For example, power generation that burns organic material from sustainable forests may be renewable, but

Email: er.chandandeep@gmail.com

Chandan Deep Singh, Talwinder Singh and Davinder Singh (eds.) Digitization and Manufacturing Performance: An Environmental Perspective, (1–50)

it is not necessarily green, due to the CO2 produced by the burning process itself.

Green energy sources are usually naturally replenished, as opposed to fossil fuel sources like natural gas or coal, which can take millions of years to develop. Green sources also often avoid mining or drilling operations that can be damaging to eco-systems.

Types

The main sources are wind energy, solar power and hydroelectric power (including tidal energy, which uses ocean energy from the tides in the sea). Solar and wind power are able to be produced on a small scale at people's homes or alternatively, they can be generated on a larger, industrial scale.

The six most common forms are as follows:

Solar Power

This common type of renewable energy is usually produced using photovoltaic cells that capture sunlight and turn it into electricity. Solar power is also used to heat buildings and for hot water as well as for cooking and lighting. Solar power has now become affordable enough to be used for domestic purposes including garden lighting, although it is also used on a larger scale to power entire neighbourhoods.

Wind Power

Particularly suited to offshore and higher altitude sites, wind energy uses the power of the flow of air around the world to push turbines that then generate electricity.

Hydropower

Also known as hydroelectric power, this type of green energy uses the flow of water in rivers, streams, dams or elsewhere to produce electricity. Hydropower can even work on a small scale using the flow of water through pipes in the home or can come from evaporation, rainfall or the tides in the oceans.

Exactly how "green" the following three types of green energy are is dependent on how they are created.

Geothermal Energy

This type of green power uses thermal energy that has been stored just under the earth's crust. While this resource requires drilling to access, thereby calling the environmental impact into question, it is a huge resource once tapped into. Geothermal energy has been used for bathing in hot springs for thousands of years and this same resource can be used for steam to turn turbines and generate electricity. The energy stored under the United States alone is enough to produce 10 times as much electricity as coal currently can. While some nations, such as Iceland, have easy-to-access geothermal resources, it is a resource that is reliant on location for ease of use, and to be fully "green" the drilling procedures need to be closely monitored.

Biomass

This renewable resource also needs to be carefully managed in order to be truly labelled as a "green energy" source. Biomass power plants use wood waste, sawdust and combustible organic agricultural waste to create energy. While the burning of these materials releases greenhouse gas these emissions are still far lower than those from petroleum-based fuels.

Biofuels

Rather than burning biomass as mentioned above, these organic materials can be transformed into fuel such as ethanol and biodiesel. Having supplied just 2.7% of the world's fuel for transport in 2010, the biofuels are estimated to have the capacity to meet over 25% of global transportation fuel demand by 2050.

Green energy is important for the environment as it replaces the negative effects of fossil fuels with more environmentally-friendly alternatives. Derived from natural resources, green energy is also often renewable and clean, meaning that they emit no or few greenhouse gases and are often readily available.

Even when the full life cycle of a green energy source is taken into consideration, they release far less greenhouse gases than fossil fuels, as well as few or low levels of air pollutants. This is not just good for the planet but is also better for the health of people and animals that have to breathe the air.

Green energy can also lead to stable energy prices as these sources are often produced locally and are not as affected by geopolitical crisis, price spikes or supply chain disruptions. The economic benefits also include job creation in building the facilities that often serve the communities where the workers are employed.

Renewable energy saw the creation of 11 million jobs worldwide in 2018, with this number set to grow as we strive to meet targets such as net zero.

Due to the local nature of energy production through sources like solar and wind power, the energy infrastructure is more flexible and less dependent on centralized sources that can lead to disruption as well as being less resilient to weather related climate change.

Green energy also represents a low cost solution for the energy needs of many parts of the world. This will only improve as costs continue to fall, further increasing the accessibility of green energy, especially in the developing world.

Examples

There are plenty of examples of green energy in use today, from energy production through to thermal heating for buildings, off-highway and transport. Many industries are investigating green solutions and here are a few examples:

Heating and Cooling in Buildings

Green energy solutions are being used for buildings ranging from large office blocks to people's homes. These include solar water heaters, biomass fuelled boilers and direct heat from geothermal, as well as cooling systems powered by renewable sources.

1.2 Industrial Processes

Renewable heat for industrial processes can be run using biomass or renewable electricity. Hydrogen is now a large provider of renewable energy for the cement, iron, steel and chemical industries.

Transport

Sustainable biofuels and renewable electricity are growing in use for transportation across multiple industry sectors. Automotive is an obvious example as electrification advances to replace fossil fuels, but aerospace and construction are other areas that are actively investigating electrification.

Can it replace fossil fuels?

Green energy has the capacity to replace fossil fuels in the future, however it may require varied production from different means to achieve this. Geothermal, for example, is particularly effective in places where this resource is easy to tap into, while wind energy or solar power may be better suited to other geographic locations.

However, by bringing together multiple green energy sources to meet our needs, and with the advancements that are being made with regards to production and development of these resources, there is every reason to believe that fossil fuels could be phased out.

We are still some years away from this happening, but the fact remains that this is necessary to reduce climate change, improve the environment and move to a more sustainable future.

Can it be economically viable?

Understanding the economic viability of green energy requires a comparison with fossil fuels. The fact is that as easily-reached fossil resources begin to run out, the cost of this type of energy will only increase with scarcity.

At the same time as fossil fuels become more expensive, the cost of greener energy sources is falling. Other factors also work in favour of green energy, such as the ability to produce relatively inexpensive localized energy solutions, such as solar farms. The interest, investment and development of green energy solutions is bringing costs down as we continue to build up our knowledge and are able to build on past breakthroughs.

As a result, green energy can not only become economically viable but also the preferred option.

Which type is the most efficient?

Efficiency in green energy is slightly dependent on location as, if you have the right conditions, such as frequent and strong sunlight, it is easy to create a fast and efficient energy solution.

However, to truly compare different energy types it is necessary to analyze the full life cycle of an energy source. This includes assessing the energy used to create the green energy resource, working out how much energy can be translated into electricity and any environmental clearing that was required to create the energy solution. Of course, environmental damage would prevent a source truly being "green," but when all of these factors are combined it creates what is known as a "Levelised Energy Cost" (LEC).

Currently, wind farms are seen as the most efficient source of green energy as it requires less refining and processing than the production of, for example, solar panels. Advances in composites technology and testing has helped improve the life-span and therefore the LEC of wind turbines. However, the same can be said of solar panels, which are also seeing a great deal of development.

Green energy solutions also have the benefit of not needing much additional energy expenditure after they have been built, since they tend to use a readily renewable source of power, such as the wind. In fact, the total efficiency of usable energy for coal is just 29% of its original energy value, while wind power offers a 1164% return on its original energy input.

Renewable energy sources are currently ranked as follows in efficiency (although this may change as developments continue):

1. Wind Power
2. Geothermal
3. Hydropower

4. Nuclear
5. Solar Power

How can it help the environment?

Green energy provides real benefits for the environment since the power comes from natural resources such as sunlight, wind and water. Constantly replenished, these energy sources are the direct opposite of the unsustainable, carbon emitting fossil fuels that have powered us for over a century.

Creating energy with a zero carbon footprint is a great stride to a more environmentally friendly future. If we can use it to meet our power, industrial and transportation needs, we will be able to greatly reduce our impact on the environment.

Green energy vs clean energy vs renewable energy—What is the difference?

As we touched upon earlier, there is a difference between green, clean, and renewable energy. This is slightly confused by people often using these terms interchangeably, but while a resource can be all of these things at once, it may also be, for example, renewable but not green or clean (such as with some forms of biomass energy).

Green energy is that which comes from natural sources, such as the sun. Clean energy are those types which do not release pollutants into the air, and renewable energy comes from sources that are constantly being replenished, such as hydropower, wind power or solar energy.

Renewable energy is often seen as being the same, but there is still some debate around this. For example, can a hydroelectric dam which may divert waterways and impact the local environment really be called "green"?

However, a source such as wind power is renewable, green and clean—since it comes from an environmentally-friendly, self-replenishing and non-polluting source [1].

The role of renewables in sustainable development

Renewable energy has an important role to play in meeting future energy needs in both rural and urban areas. The development and utilization of renewable energy should be given a high priority, especially in the light of increased awareness of the adverse environmental impacts of fossil-based generation. The need for sustainable energy development is increasing rapidly in the world. Widespread use of renewable energy is important for achieving sustainability in the energy sectors in both developing and industrialized countries.

Renewable energy resources and technologies are a key component of sustainable development for three main reasons:

- They generally cause less environmental impact than other energy sources. The variety of renewable energy resources provides a flexible array of options for their use.
- They cannot be depleted. If used carefully in appropriate applications, renewable energy resources can provide a reliable and sustainable supply of energy almost indefinitely. In contrast, fossil fuel and uranium resources are diminished by extraction and consumption.
- They favor system decentralization and local solutions that are somewhat independent of the national network, thus enhancing the flexibility of the system and providing economic benefits to small isolated populations. Also, the small scale of the equipment often reduces the time required from initial design to operation, providing greater adaptability in responding to unpredictable growth and/or changes in energy demand.

Environment and renewable energy technologies

Renewable energy technologies" is an umbrella term that stands for energy production using a renewable energy source like solar, wind, water (hydro and tidal), biomass (biofuels and wastes), and geothermal heat.

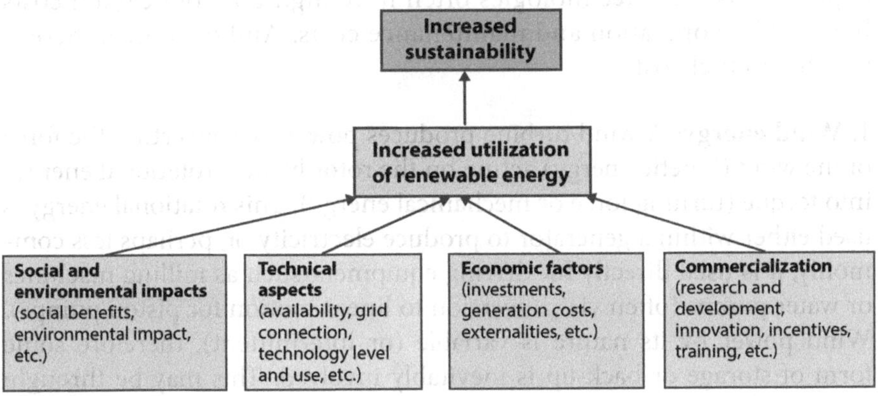

This module provides an outline and brief description, including fundamentals, of the different renewable energy technologies, wind, solar, bioenergy, hydro and geothermal energy. It provides a general overview of the technologies and their applications. Electricity generation from wave and tidal energy is not discussed.

The use of this technology is less relevant for developing countries as mostly these technologies are still at the prototype stage. While these technologies are not fully proven yet, promising research and development is being conducted. The module also reviews the costs of the different technologies and discusses common technical and non-technical barriers and issues limiting the wide spread use/dissemination of renewable energy in developing countries. The information in this module is of general interest to explain the basics of renewable energy technologies, to understand their strengths and weaknesses and hence to have a better grasp of the benefits available from, and the barriers faced by, these technologies.

1.3 Overview of Renewable Energy Technologies

One of the first aspects to consider is the cost of renewable energy technologies. However, this is not an easy question to answer because, as with many energy technologies, many factors affect cost and different sources of information use different criteria for estimating cost. In many cases, the environmental benefits of renewable energy technologies are difficult to take into account in terms of cost savings through less pollution and less damage to the environment. When trying to calculate the cost of these technologies is often best to take a life cycle cost approach, as these technologies often have high up-front capital costs but very low operation and maintenance costs. And of course, there is usually no fuel cost.

1. **Wind energy:** A wind turbine produces power by converting the force of the wind (kinetic energy) acting on the rotor blades (rotational energy) into torque (turning force or mechanical energy). This rotational energy is used either within a generator to produce electricity or, perhaps less commonly, it is used directly for driving equipment such as milling machines or water pumps (often via conversion to linear motion for piston pumps). Wind power by its nature is variable (or intermittent), therefore some form of storage or back-up is inevitably involved. This may be through: (a) connection to an electricity grid system, which may be on a large or small (mini-grid) scale; incorporating other electricity producing energy systems (from conventional generating stations through diesel generators to other renewable energy systems).

Wind turbines generating electricity

Several turbine types exist but presently the most common configuration has become the horizontal axis three bladed turbine. The rotor may be positioned up or downwind (although the former is probably the most common). Modern wind turbines vary in size with two market ranges: small units rated at just a few hundred watts up to 50-80 kW in capacity, used mainly for rural and stand-alone power systems; and large units, from 150 kW up to 5 MW in capacity, used for large-scale, grid-connected systems.

Grid-connected wind turbines

Grid-connected wind turbines are certainly having a considerable impact in developed countries and in some developing countries, namely Argentina, China and India. This is mainly through large-scale installations either on land (on-shore) or in the sea on the continental shelf (off-shore). In addition, in developed countries, more smaller machines are now being grid-connected. These are usually installed to supply power to a private owner already connected to the electricity grid but who wishes to supply at least some of their own power. This principle can be used in developing countries to contribute to a more decentralized grid network and/or to support a weak grid [2].

Stand-alone wind turbines

The most common type of stand-alone small wind electric system involves the use of a wind generator to maintain an adequate level of charge in an electrical storage battery. The battery in turn can provide electricity on demand for electrical applications such as lights, radios, refrigeration, telecommunications, etc., irrespective of whether or not the wind is blowing. A controller is also used to ensure that the batteries are not damaged by overcharging (when surplus energy is dissipated through a dump load) or excessive discharge, usually by sensing low voltage. Loads connected to the battery can either be DC or AC (via an inverter). Small wind battery charging systems are most commonly rated at between 25 and 100 W for a 10m/s wind speed, and are quite small with a rotor diameter of 50 cm to 1 m. These systems are suitable for remote settlements in developing countries. Larger stand-alone systems, incorporating larger wind electricity generators and correspondingly larger battery banks (at an increased cost) are also available, these may include other renewable energy technologies, such as PV, as well as diesel generators to ensure that the batteries are always charged and that power availability is high. Less common is the stand-alone system which does not incorporate a battery back. This involves the use of a wind turbine with, at least, a diesel generator, which will automatically supply power when required. This has the advantage of not requiring a battery bank but the required control systems are complex.

Wind turbines for water pumping

The most common type of a mechanical wind turbine is the wind pump which uses the wind's kinetic energy to lift water. Wind pumps are typically used for water supply (livestock or human settlements), small-scale irrigation or pumping seawater for sea salt production. Here we look at the two main uses which are irrigation and water supply. There are two distinct categories of wind pump, because the technical, operational and economic requirements are generally different for these two end uses. That is not to say that a water supply wind pump cannot be used for irrigation (they quite often are) but irrigation designs are generally unsuitable for water supply duties.

Most water supply wind pumps must be ultra-reliable, to run unattended for most of the time (so they need automatic devices to prevent over-speeding in storms), and they also need the minimum of maintenance and attention and to be capable of pumping water generally from depths of 10 m to 100 m or more. A typical farm wind pump should run for over 20 years with maintenance only once every year, and without any major replacements; this is a very demanding technical requirement since typically such a wind pump must average over 80,000 operating hours before anything significant wears out; this is four to ten times the operating life of most small diesel engines or about 20 times the life of a small engine pump. Wind pumps to this standard therefore are usually industrially manufactured from steel components and drive piston pumps via reciprocating pump rods. Inevitably they are quite expensive in relation to their power output, because of the robust nature of their construction. But American, Australian and Argentine ranchers have found the price worth paying for wind pumps that run and run without demanding much attention to the extent they can almost be forgotten about for weeks at a time. This inherent reliability for long periods is their main advantage over practically any other form of pumping system. Irrigation duties on the other hand are seasonal (so the windmill may only be useful for a limited fraction of the year), they involve pumping much larger volumes of water through a low head, and the intrinsic value of the water is low when compared with drinking water. Therefore any wind pump developed for irrigation has to be as cheap as possible and this requirement tends to override most other considerations. Since irrigation generally involves the farmer and/or other workers being present, it is not so critical to have a machine capable of running unattended. Therefore windmills used for irrigation in the past tend to be indigenous designs that are often improvised or built by the farmer as a method of low-cost mechanization. Most farm wind pumps, even though still in commercial production, date back to the 1920s or earlier and are therefore heavy and expensive to manufacture, and difficult to install properly in remote areas. Recently, various efforts have been made to revise the traditional farm wind pump concept into a lighter and simpler modern form. Modern designs are fabricated from standard steel stock by small engineering companies and cost (and weigh) only about half as much as traditional American or Australian machines of similar capability. It is possible therefore that through developments of this kind, costs might be kept low enough to allow the marketing of all-steel-wind-pumps that are both durable, like the traditional designs, yet cheap enough to be economic for irrigation.

2. **Solar energy:** Solar energy technologies can be loosely divided into two categories: solar thermal systems and solar electric or photovoltaic (PV) systems. Photovoltaic (PV) systems Photovoltaic or PV devices convert sun light directly into electrical energy. The amount of energy that can be produced is directly dependent on the sunshine intensity. Thus, for example, PV devices are capable of producing electricity even in winter and even during cloudy weather albeit at a reduced rate. Natural cycles in the context of PV systems thus have three dimensions. As with many other renewable energy technologies, PV has a seasonal variation in potential electricity production with the peak in summer although in principle PV devices operating along the equator have an almost constant exploitable potential throughout the year. Secondly, electricity production varies on a diurnal basis from dawn to PV devices use the chemical-electrical interaction between light radiation and a semiconductor to obtain DC electricity. The base material used to make most types of solar cell is silicon (approx. 87%). The main technologies in use today are: Mono-crystalline silicon cells are made of silicon wavers cut from one homogenous crystal in which all silicon atoms are arranged in the same direction. These have a conversion efficiency of 12–15%. Poly-crystalline silicon cells are poured and are cheaper and simpler to make than mono crystalline silicon and the efficiency is lower than that of monocrystalline cells (conversion efficiency 11–14%).

3. Thin film solar cells are constructed by depositing extremely thin layer of photovoltaic materials on a low-cost backing such as glass, stainless steel or plastic (conversion efficiency 5–12%); Multiple junction cells use two or three layers of different materials in order to improve the efficiency of the module by trying to use a wider spectrum of radiation (conversion efficiency

20–30%). The building block of a PV system is a PV cell. Many PV cells are encapsulated together to form a PV panel or module. A PV array, which is the complete powergenerating unit, consists of any number of PV modules/panels. Depending on their application, the system will also require major components such as a battery bank and battery controller, DC-AC power inverter, auxiliary energy source etc. Individual PV cells typically have a capacity between 5 and 300 W but systems may have a total installed capacity ranging from 10 W to 100 MW. The very modular nature of PV panels as building blocks to a PV system gives the sizing of systems an important flexibility.

Solar thermal systems: Solar thermal systems use the sun's power in terms of its thermal or heat energy for heating, drying, evaporation and cooling. Many developing countries have indigenous products such as solar water heaters, solar grain dryers, etc. These are usually local rather than international products, specific to a country or even to a region. The main solar thermal systems employed in developing countries are discussed briefly below.

Solar thermal power plants: Solar thermal engines use complex concentrating solar collectors to produce high temperatures. These temperatures are high enough to produce steam, which can be used to drive steam turbines generating electricity. There is a wide variety of different designs, some use central receivers (where the solar energy is concentrated to a tower) whilst others use parabolic concentrator systems. Although the first commercial thermal power plants have been in operation in California since the mid-1980s, many of the newer designs are still at the prototype stage being tested in pilot installations in the deserts of the United States and elsewhere. The Global Environment Facility (GEF) has supported the first planning phase of a project that is developing a concentrating solar power plant in Egypt in 2004. There are also projects in India, Mexico and Morocco that have been supported by GEF as part of a strategy to accelerate cost reduction and commercial adoption of high temperature solar thermal energy technology.

Solar water heating: Solar water heating system may be used in rural clinics, hospitals or even schools. The principle of the system is to heat water, usually in a special collector and store it in a tank until required. Collectors are designed to collect the heat in the most efficient, but cost effective way, usually into a heat transfer fluid, which then transfers its heat to the water in the storage tank. The two main types of collector are: flat plate and evacuated tube. For example, to heat 100 L of water through a temperature rise of 40°C with a simple flat plate solar collector requires only approximately 2.5 m2 of collector area but saves approx. 10 kg of woodfuel that would normally be required to heat this quantity of water. The cheapest technology available and the simplest to install is a thermosiphon system, which uses the natural tendency of heated water to rise and cooler water to fall to perform the heat collection task, water inside the collector flow-tubes is heated. As it heats, this water expands slightly and becomes lighter than the cold water in the solar storage tank mounted above the collector. Gravity then pulls the heavier, cold water down from the tank and into the collector inlet. The cold water pushes the heated water through the collector outlet and into the top of the tank, thus heating the water in the tank.

Solar drying: Solar drying, in the open air, has been used for centuries. Drying may be required to preserve agricultural/food products or as a part of the production process, i.e. timber drying. Solar drying systems are those that use the sun's energy more efficiently than simple open-air drying.

In general, solar drying is more appropriate when: The higher the value per ton of products dried; The higher the proportion of the product currently spoiled in the open air; The more often the drier will be used.

Solar cookers: Solar cookers can be important because of the increased scarcity of wood fuel and the problems of deforestation in many developing country regions. Solar cookers can also promote cleaner air where there is a problem with indoor cooking. There are basically two types of solar cooker: oven or stove type. As with conventional cooking stoves, solar stoves apply heat to the bottom of the cooking pot while solar ovens apply a general heat to the enclosed area which contains the cooking pot. However, there are important social issues related to the effective use of solar cookers. There will always be some change of habits required and readiness to change is an important factor that affects the potential impact of this technology.

Solar distillation: Solar distillation is a solar enhanced distillation process to produce potable water from a saline source. It can be used in areas where, for instance, drinking water is in short supply but brackish water, i.e. containing dissolved salts, is available. In general solar distillation equipment, or stills, is more economically attractive for smaller outputs. Costs increase significantly with increased output, in comparison to other technologies which have considerable economics of scale.

Solar cooling: Several forms of mature technologies are available today for solar-thermally assisted air-conditioning and cooling applications. In particular for centralized systems providing conditioned air and/or chilled water to buildings, all necessary components are commercially available.

The great advantage of this solar application, especially in tropical and equatorial countries, is that the daily cooling load profile follows the solar radiation profile (i.e. office buildings).

Bioenergy: Bioenergy is a general term that covers energy derived from a wide variety of material of plant or animal origin. Strictly, this includes fossil fuels but, generally, the term is used to mean renewable energy sources such as wood and wood residues, agricultural crops and residues, animal fats, and animal and human wastes, all of which can yield useful fuels either directly or after some form of conversion. There are technologies for bioenergy using liquid and gaseous fuel, as well as traditional applications of direct combustion. The conversion process can be physical (for example, drying, size, reduction or densification), thermal (as in carbonization) or chemical (as in biogas production). The end result of the conversion process may be a solid, liquid or gaseous fuel and this flexibility of choice in the physical form of the fuel is one of the advantages of bioenergy over other renewable energy sources. The basis for all these applications is organic matter, in most cases plants and trees. There is a trend towards purposefully planted biomass energy crops, although biomass can also be collected as a by-product and residue from agricultural, forestry, industry and household waste. Bioenergy can be used for a great variety of energy needs, from heating and transport fuel to power generation. There are numerous commercially available technologies for the conversion processes and for utilization of the end-products. Although the different types of bioenergy have features in common, they exhibit considerable variation in physical and chemical characteristics which influence their use as fuels. There is such a wide range of bioenergy systems that this module does not aim to cover and describe each one.

4. **Hydro:** Hydropower is the extraction of energy from falling water (from a higher to a lower altitude) when it is made to pass through an energy conversion device, such as a water turbine or a water wheel. A water turbine converts the energy of water into mechanical energy, which in turn is often converted into electrical energy by means of a generator. Alternatively, hydropower can also be extracted from river currents when a suitable device is placed directly in a river. The devices employed in this case are generally known as river or water current turbines1 or a "zero head" turbine. This module will review only the former type of hydropower, as the latter has a limited potential and application. Hydropower systems can range from tens of Watts to hundreds of Megawatts. However, there is no internationally recognized standard definition for hydropower sizes, so definitions can vary from one country to another. On a smaller scale, used more often in

rural and remote areas, micro-hydro schemes can have capacities up to 500 kW and are generally run-of-the-river developments for villages. On an even smaller scale pico-hydro systems tend to be between 50 W and 5 kW and are generally used for individual homes or clusters of households.

Hydropower, under the right circumstances, can be one of the most reliable and cost-effective renewable energy sources. The applications of small-hydro facilities include base, peak and stand-by power production or stand-alone applications. Hydroelectric plants typically generate power between 15% and 100% of the time. In base loading applications, units must be able to operate at least 85% of the time. SHP installations commonly last without the need for major replacement costs for 30+ years. Within the limits of water resources available, SHP installations are characterized by reliability and flexibility of operation, including fast start-up and shut-down in response to rapid demand changes. SHP electricity can be tailored to the needs of the end-use market, avoiding balance and power reliability concerns. Hydropower, under the right circumstances, can be one of the most reliable and cost-effective renewable energy sources. The applications of small- hydro facilities include base, peak and stand-by power production or stand-alone applications. Hydroelectric plants typically generate power between 15% and 100% of the time. In base loading applications, units must be able to operate at least 85% of the time. SHP installations commonly last without the need for major replacement costs for 30+ years. Within the limits of water resources available, SHP installations are characterized by reliability and flexibility of operation, including fast start-up and shut-down in response to rapid demand changes. SHP electricity can be tailored to the needs of the end-use market, avoiding balance and power reliability concerns.

SHP does not have the same kinds of adverse effects on the local environment as large hydro. Nevertheless, SHP has some adverse impacts on the environment. For example when water levels in reservoirs change abruptly to meet electricity demands or in times of low flow, the short stretch of by-passed river can run dry, which might dry out aquatic organisms. Power plants often obstruct the natural migration of fish through the river system. Such effects could result in the extinction of fish populations, a fundamental change of natural flow regimes, the loss of aquatic habitats, sinking groundwater levels and a deterioration of landscapes. In order for hydropower plants to be socially and environmentally sustainable, the local and regional impacts need to be evaluated, reduced and minimized. 3 Therefore, modern construction designs typically implement mitigation measures. The Eugene Standard4 provides a set of criteria that hydropower plants (regardless of their installed capacity, age or mode of operation) need to comply with in order to be an environmentally sound form of power supply.

Under the Standard, three basic conditions have to be met: 1. The basic requirements include (non-exhaustive list): (a) Hydropower utilization should not lastingly impair the ability of fish to migrate unimpeded through the affected river system. (b) Hydropower utilization should not result in long-term degradation in the natural diversity of plants and animals. (c) Power plant constructions should not irreversibly destroy protected habitats. (d) Power plant operation should not endanger fish or benthic organisms in affected river reaches. 2. Secondly the sustainable hydropower plant should invest money (for instance a fixed rate per kilowatt-hour produced or sold) to restore, protect or upgrade the environment surrounding the hydropower plant (so-called ecoinvestments). 3. Thirdly, to assess to which extent the basic requirements were met and what the eco-investments were used for, an audit and certification procedure should be carried out every year [2].

5. **Geothermal:** Geothermal is energy available as heat emitted from within the earth, usually in the form of hot water or steam. Geothermal heat has two sources: the original heat produced from the formation of the earth by gravitational collapse and the heat produced by the radioactive decay of various isotopes. It is very site dependent as the resource needs to be near surface and can be used for heating and power generation purposes. High temperature resources (150° C+) can be used for electricity generation, while low temperature resources (50-150° C) can be used for various direct uses such as district heating and industrial processing. Since the earth's crust is continuously emitting heat towards its surface at a rate of 40 million megawatts, geothermal is in principle an inexhaustible energy source, with the centre of the earth having cooled down by only about 2% over the earth's lifetime of about 4 billion years.

The extraction of energy from geothermal aquifers uses naturally occurring ground water from deep porous rocks. Water can be extracted via a production borehole and, generally be disposed of via an injection hole. Another method is the extraction of heat from hot dry rock (HDR) which uses reservoirs created artificially by hydraulic fracturing. Heat is extracted by circulating water under pressure via production wells. There are no problems of intermittency in the utilization of geothermal energy sources for direct heat applications or for electricity generation. A developed geothermal field provides what is essentially a distributed heat source, since the input to a power plant normally consists of the integrated outputs of several wells. Thus one or more wells may be shut for repairs or maintenance while others produce. Proper dimensioning of the generating plant ensures that there is always enough steam or hot water available for operation. This feature and the low operational costs are the reasons why geothermal power plants are normally used for base load power. Natural variations of geothermal resources occur over extremely long periods, millennia or even longer time scales. However, man-induced processes lead to variations with shorter time scales, typically in the range of decades. Unwanted effects of overexploitation and improper reinjection have been observed, especially in the early years of geothermal technology development. However, presentday geothermal technology for field characterization and modelling makes it possible to avoid improper practices, or at worst to detect their effects at an early stage before they become significant. Environmentally, geothermal schemes are relatively benign, but typically do produce a highly corrosive brine which may need special treatment and discharge consents. There is also a possibility of noxious gases, such as hydrogen sulphide, being emitted.

Bio-energy & Conversion Systems: Biomass conversion is a shared area between hydrogen production and biogas production. It is similar to coal gasification in terms of converting the original resource to a hydrogen-containing gas at high temperatures without combustion. The major options within thermochemical biomass conversion processes include combustion, gasification, pyrolysis, and liquefaction. The most practiced thermochemical conversion of biomass industrially is combustion process, which is used for heat and electricity generation. The major commercial fuels used in the world today are natural gas, gasoline (petrol), aviation fuel, diesel, fuel oils, and solid fuels such as coal. These commercial fossil fuels could be replaced with biofuels and solid fuels derived from biomass by using conversion technologies. There are specific biomass resources that are well-suited to each conversion technology. For example, sugar crops (sugarcane and sweet sorghum) are good feedstock materials for the conversion of bioethanol; oil crops (soybean and canola oil) are ideal feedstock for biodiesel production; and lignocellulosic biomass (e.g., wood wastes, animal manure or grasses) is the prime substrate for making biogas. Thermal conversion systems convert all other biomass resources into valuable products.

Replacement of these primary fuels with bio-based alternatives is one way to address energy sustainability. Heat and electrical power, needed worldwide, can also be produced through the conversion of biomass through thermo-chemical conversion processes such as pyrolysis and gasification to produce synthesis gas (or also called syngas, a shorter version).

Syngas can be combusted to generate heat and can be thoroughly cleaned of tar and used in an internal combustion engine to generate mechanical or electrical power. Future world requirements for other basic energy and power needs can be met using a wide range of biomass resources, including oil and sugar crops, animal manure, crop residues, municipal solid wastes (MSW), fuel wood, aquatic plants like micro-algae, and dedicated energy farming for energy production. The three primary products of thermal conversion are solid bio-char, liquid, and synthesis gas. Replacement of these primary fuels with bio-based alternatives is one way to address energy sustainability. Heat and electrical power, needed worldwide, can also be produced through the conversion of biomass through thermo-chemical conversion processes such as pyrolysis and gasification to produce synthesis gas (or also called syngas, a shorter version). Syngas can be combusted to generate heat and can be thoroughly cleaned of tar and used in an internal combustion engine to generate mechanical or electrical power. Future world requirements for other basic energy and power needs can be met using a wide range of biomass resources, including oil and sugar crops, animal manure, crop residues, municipal solid wastes (MSW), fuel wood, aquatic plants like micro-algae, and dedicated energy farming for energy production. The three primary products of thermal conversion are solid bio-char, liquid, and synthesis gas.

Biodiesel Production:
Refined vegetable oils and fats are converted into *biodiesel*, which is compatible with diesel fuel, by physicochemical conversion using a simple catalytic process using methanol (CH3OH) and sodium hydroxide (NaOH) at a slightly elevated temperature. The process is called transesterification. Vegetable oils are also called triglycerides because their chemical structure is composed of a glycerol attached to three fatty acid molecules by ester bonds. When the ester bonds are broken by a catalyst, glycerin is produced and the fatty acid compound is converted into its methyl ester form, which is the technical term for biodiesel. The combination of methanol and sodium hydroxide results in a compound called sodium methoxide (CH3ONa), which is the most common commercial catalyst for biodiesel production. The basic mass balance for the process is:

$$100\text{ kg vegetable oil} + 10\text{ kg catalysts} \rightarrow 100\text{ kg biodiesel} + 10\text{ kg glycerin}$$

The energy balance depends on the specific facility design. For the biodiesel product to be considered viable, the energy in the biodiesel must exceed the energy used to produce the vegetable oil used for the process. In a commercial system, the transesterification process is split into several stages. Methanol and catalysts are recovered after all the stages to minimize catalyst consumption. Crude glycerin is also recovered at each stage to minimize the use of excess methanol. The remaining catalyst—the amount must be calculated accurately—is then introduced at the last stage of the process. This last stage reaction minimizes unreacted mono-glycerides (or remaining glycerol that still has a fatty acid chained to it via an ester bond). If soybean oil is used, the resulting biodiesel product is called soybean methyl ester, the most common biodiesel product in the United States. In Europe, canola (rapeseed) oil is the most common feedstock, which produces rapeseed methyl ester. The glycerin co-product is further purified to improve its commercial value.

Bioethanol Production:

Bioethanol, which is compatible with gasoline or petrol, is produced from sugar, starchy, or lignocellulosic crops using microbes or enzymes. Sugars from crops are easily converted into ethanol using yeast (e.g., *Saccharomyces cerevisiae*) or other similar microbes, while starchy crops need enzymes (e.g., amylases) that convert starch to sugar, with the yeasts then acting on the sugars to produce bioethanol. Lignocellulosic crops need similar enzymes (e.g., enzymes produced by *Trichoderma reesei*) to break down cellulose into simple sugars. The basic mass balance for the conversion of plant sugars from biomass into ethanol (C2H6O) also yields heat:

C6H12O6+ yeast →2C2H6O+2CO2(+ heat)C6H12O6+ yeast → 2C2H6O+2CO2(+ heat)

The most common feedstock for making bioethanol in the United States is dry milled corn (maize; *Zea mays*). In the process, dry corn kernels are milled, then water is added to the powdered material while being heated (or gelatinized) in order to cook the starch and break it down using the amylase enzyme (saccharification). This process converts starch into sugars. The resulting product (mainly glucose) is then converted into bioethanol using yeasts fermentation for 3-5 days with a mass balance of:

2C6H10O5+H2O+ amylase →C12H22O112C6H10O5+H2O+

amylase →C12H22O11 or

$$C12H22O11+H2O+ \text{invertase} \rightarrow 2C6H12O6C12H22O11+H2O+$$

$$\text{invertase} \rightarrow 2C6H12O6$$

In this representation, complex starch molecules are represented by repeating units of polymers of glucose [(C6H10O5)n] with n being any number of chains. The enzyme amylase reduces this polymer into simple compounds, such as sucrose (C12H22O11), a disaccharide having just two molecules of glucose. Alternatively, the enzyme invertase is used to break down sucrose into glucose sugar. A yeast, such as the commercial yeast Ethanol Red (distributed by Fermentis of Lesaffre, France and sold worldwide) acts on the sugar product to convert the sugar into bioethanol. The resulting product (a broth) is called *beer* because its alcohol content is very close to 10%. The solid portion is called *distillers grain*, which is usually dried and fed to animals. The beer is distilled to yield solids (known as bottoms or still bottoms) and to recover 90-95% of the bioethanol (usually 180-190 proof), which is then purified using molecular sieves. (A molecular sieve is a crystalline substance with pores of carefully selected molecular dimensions that permit the passage of, in this case, only ethanol molecules.) The final separated and purified product may then be blended with gasoline or used alone.

Biogas Production:
Biogas, which is composed primarily of methane (CH4; also called natural gas) and carbon dioxide (CO2), is produced from lignocellulosic biomass by microbes under anaerobic conditions. Suitable microbes are commonly found in the stomachs of ruminant animals (e.g., cows). These microbes convert complex cellulosic materials into organic acids via hydrolysis or fermentation; these large organic acids are further converted into simpler organic acids (e.g., acetic acids) and hydrogen gas. Hydrogen gas and some organic acids that include CO2 are further converted into CH4 and CO2 as the respiratory gases of these microbes. Biogas (CO2 + CH4) is the same as natural gas (CH4) if the CO2 component is removed. Natural gas is a common fuel derived by refining crude oil.

Biomass Pyrolysis:
Pyrolysis is a thermal conversion process at elevated temperatures in complete absence of oxygen or an oxidant. The primary products are solid bio-char, liquid, and gaseous synthesis gas. The ratios of these co-products depend on temperature, retention time, and type of biomass used. The quality and magnitude of products are also dependent on the reactor used. The simple rules of biomass pyrolysis processes are:

1. Solid bio-char (or charcoal) yield is maximized at the lowest pyrolysis temperature and the longest residence time.
2. Liquid yield is usually maximized at temperatures between 400°C and 600°C.
3. Synthesis gas, or syngas, is maximized at the highest operating temperature. The main components of syngas are carbon monoxide (CO) and hydrogen (H2). Other component gases include lower molecular weight hydrocarbons such as CH4, ethylene (C2H4), and ethane (C2H6).

Economic Evaluation of Bioenergy Systems:
Commercial bioenergy facilities depreciate every year. There is no accurate estimate of depreciation values but a potential investor may use this parameter to save on capital costs each year from the proceeds of the commercial facility such that at the end of the life of the facility, the investor is prepared to invest in higher-yielding projects.

There are a number of simple methods that engineers may use for economic depreciation analyses of bioenergy facilities. A basic economic evaluation is required early in the design of the system to ascertain feasibility prior to significant capital investment. Evaluation of the economic feasibility begins with the analysis of the *fixed* (or capital) expenditures and *variable* (or operating) costs (Watts and Hertvik, 2018). Fixed expenditures include the capital cost of assets such as biomass conversion facilities, land, equipment, and vehicles, as well as depreciation of facilities and equipment, taxes, shelter, insurance, and interest on borrowed money. Variable costs are the daily or monthly operating costs for the production of a biomass product. Variable costs are associated with feedstock and chemicals, repair and maintenance, water, fuel, utilities, energy, labor, management, and waste disposal. Fixed costs do not vary with time and output while variable costs increase with time and output of product. The total project cost is the sum of fixed and variable costs. Variable costs per unit of output decrease with increased amount of output, so the profitability of a product may depend on the amount produced [3].

Sustainability Issues in Biomass Energy Conversion Systems:
The US Department of Energy (USDOE) and US Department of Agriculture (USDA) define sustainable biofuels as those that are "economically competitive, conserve the natural resource base, and ensure social well-being." The conservation of resource base points to the conservation of energy as well, that is, the fuel produced must have more energy than the total energy used to produce the fuel. One of the most common indicators of sustainability for biomass utilization is energy use throughout the life

cycle of production. There are two measures used for this evaluation: the net energy ratio (NER) and net energy balance (NEB) NER must be greater than 1 and NEB must be positive for the system to be considered sustainable from an energy perspective.

Applications:
Engineers assigned to design, operate, and manage a commercial biodiesel plant must decide what working system to adopt. The cheapest and most common is the use of gravity for separating the biodiesel (usually the top layer) and glycerin (the bottom layer). An example of this commercial operational facility is the 3 million gallon per year (MGY) (11.36 ML/yr) biodiesel plant in Dayton, Texas, operated by AgriBiofuels, LLC. This facility began operation in 2006 and is still in operation. The biodiesel recovery for this facility is slightly lower than those with computer-controlled advanced separation systems using centrifuges. Thus, one would expect their conversion efficiency and biodiesel recovery to be lower.

Biodiesel production is an efficient biomass conversion process. The ideal mass balance equation, presented earlier, is:

$$100 \text{ kg vegetable oil} + 10 \text{ kg catalysts} \rightarrow 100 \text{ kg biodiesel} + 10 \text{ kg glycerin}$$

The relationship shows that an equivalent mass of biodiesel is produced for every unit mass of vegetable oil used, but there are losses along the way and engineers must consider these losses when designing commercial facilities. In a commercial biodiesel facility, the transesterification process is split into several reactors However, to save on capital costs, some plant managers simply divide the process into two stages. Separating glycerin and biodiesel fuel is also an issue that the engineer will be faced with. Efficient separation systems that use centrifuges are expensive compared with physical separation, and this affects the overall economy of the facility. If the initial capital available is limited, investors will typically opt for cheaper, physical gravity separation instead of using centrifuges. Crown Iron Works (in Blaine, MN) sells low cost biodiesel facilities that employ gravity separation while GEA Wesfalia (Oelde, Germany) sell more expensive biodiesel facilities that use separation by centrifuge. The latter, expensive, system is more efficient at separating glycerin and biodiesel fuel and may be beneficial in the long term, allowing the facility to sell glycerin products with minimal contamination. The engineer may compare these systems in terms of costs and efficiencies. This means the engineer must determine the agricultural land area required both for the facility and the supply of biomass. There are standard tables of oil yields from crops that are used.

Over the last century, there has been increasing debate concerning the use of biomass for different purposes such as foods, feeds, energy fuels, heating, cooling and most importantly biorefinery feedstock. The biorefinery products were aimed to replace fossil fuels and chemicals as they are renewable form of energy. Biomass is a biodegradable product from agricultural wastes and residues, forestry and aquaculture. Biomass could be sourced from a variety of raw materials such as wood and wood processing by-products, manure, fractions of organic waste products and agricultural crops. As a form of renewable energy, they have the advantages of easy storage, transportation, flexible load utilization and versatile applications. The aim of this study is to provide an overview for thermochemical and biochemical biomass conversion technologies that were employed currently. Attention was also paid to manufacture of biofuels because of their potentials as key market for large-scale green sustainable biomass product [4]. Figure 1.1 shows the schematics nested renewal fuels categories under the renewable fuels standards in USA.

Biodiesel: A fuel derived from organic oils, such as vegetable oil, rather than petroleum. Biodiesel's use and production are increasing. It is typically used for aircraft, vehicles and as heating oil. Vegetable oils comprise

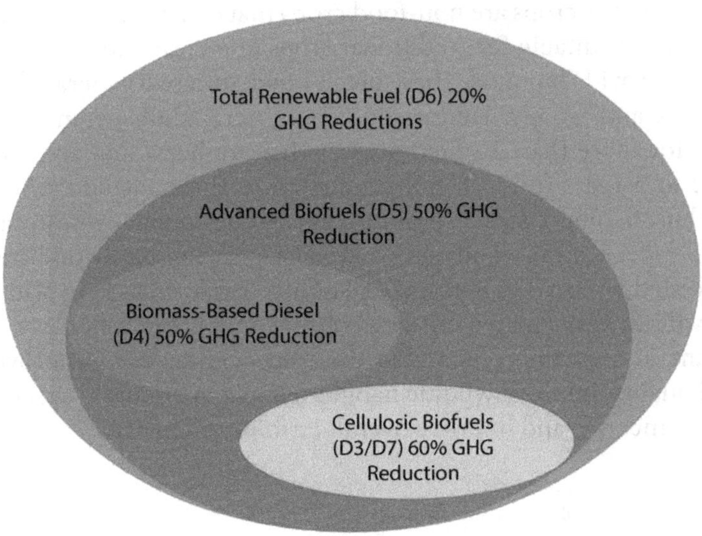

Figure 1.1 Schematic of USA nested renewable fuels categories under the renewable fuels standards.

of 90% to 95% triglycerides with small amount of diglycerides, free fatty acids, phospholipids, etc. The viscosity of vegetable oils are higher and their molecular weights are in the range of 600 to 900, which are about 3 times higher than those of the diesel fuels [5].

Biomass Feedstock:
Biomass feedstocks include dedicated energy crops, agricultural crop residues, forestry residues, algae, wood processing residues, municipal waste, and wet waste (crop wastes, forest residues, purpose-grown grasses, woody energy crops, algae, industrial wastes, sorted municipal solid waste [MSW], urban wood waste. Biomass energy supports U.S. agricultural and forest-product industries. The main biomass feedstocks for power are paper mill residue, lumber mill scrap, and municipal waste. For biomass fuels, the most common feedstocks used today are corn grain (for ethanol) and soybeans (for biodiesel). Biomass energy supports U.S. agricultural and forest-product industries. The main biomass feedstocks for power are paper mill residue, lumber mill scrap, and municipal waste. For biomass fuels, the most common feedstocks used today are corn grain (for ethanol) and soybeans (for biodiesel).

1.4 Dedicated Energy Crops

Dedicated energy crops are non-food crops that can be grown on marginal land (land not suitable for traditional crops like corn and soybeans) specifically to provide biomass. These break down into two general categories: herbaceous and woody. Herbaceous energy crops are perennial (plants that live for more than 2 years) grasses that are harvested annually after taking 2 to 3 years to reach full productivity. These include switchgrass, miscanthus, bamboo, sweet sorghum, tall fescue, kochia, wheatgrass, and others. Short-rotation woody crops are fast-growing hardwood trees that are harvested within 5 to 8 years of planting. These include hybrid poplar, hybrid willow, silver maple, eastern cottonwood, green ash, black walnut, sweetgum, and sycamore. Many of these species can help improve water and soil quality, improve wildlife habitat relative to annual crops, diversify sources of income, and improve overall farm productivity.

1.5 Agricultural Crop Residue

There are many opportunities to leverage agricultural resources on existing lands without interfering with the production of food, feed, fiber, or forest products. Agricultural crop residues, which include the stalks and leaves, are abundant, diverse, and widely distributed across the United States. Examples include corn stover (stalks, leaves, husks, and cobs), wheat straw, oat straw, barley straw, sorghum stubble, and rice straw. The sale of these residues to a local biorefinery also represents an opportunity for farmers to generate additional income.

1.6 Forestry Residues

Forest biomass feedstocks fall into one of two categories: forest residues left after logging timber (including limbs, tops, and culled trees and tree components that would be otherwise unmerchantable) or whole-tree biomass harvested explicitly for biomass. Dead, diseased, poorly formed, and other unmerchantable trees are often left in the woods following timber harvest. This woody debris can be collected for use in bioenergy, while leaving enough behind to provide habitat and maintain proper nutrient and hydrologic features. There are also opportunities to make use of excess biomass on millions of acres of forests. Harvesting excessive woody biomass can reduce the risk of fire and pests, as well as aid in forest restoration, productivity, vitality, and resilience. This biomass could be harvested for bioenergy without negatively impacting the health and stability of forest ecological structure and function.

1.7 Algae

Algae as feedstocks for bioenergy refers to a diverse group of highly productive organisms that include microalgae, macroalgae (seaweed), and cyanobacteria (formerly called "blue-green algae"). Many use sunlight and nutrients to create biomass, which contains key components—including lipids, proteins, and carbohydrates—that can be converted and upgraded to a variety of biofuels and products. Depending on the strain, algae can grow by using fresh, saline, or brackish water from surface water sources, groundwater, or seawater. Additionally, they can grow in water from second-use sources, such as treated industrial wastewater; municipal, agricultural, or aquaculture wastewater; or produced water generated from oil and gas drilling operations.

1.8 Wood Processing Residues

Wood processing yields byproducts and waste streams that are collectively called wood processing residues and have significant energy potential. For example, the processing of wood for products or pulp produces unused sawdust, bark, branches, and leaves/needles. These residues can then be converted into biofuels or bioproducts. Because these residues are already collected at the point of processing, they can be convenient and relatively inexpensive sources of biomass for energy.

1.9 Sorted Municipal Waste

MSW resources include mixed commercial and residential garbage, such as yard trimmings, paper and paperboard, plastics, rubber, leather, textiles, and food wastes. MSW for bioenergy also represents an opportunity to reduce residential and commercial waste by diverting significant volumes from landfills to the refinery.

1.10 Wet Waste

Wet waste feedstocks include commercial, institutional, and residential food wastes (particularly those currently disposed of in landfills); organic-rich biosolids (i.e., treated sewage sludge from municipal wastewater); manure slurries from concentrated livestock operations; organic wastes from industrial operations; and biogas (the gaseous product of the decomposition of organic matter in the absence of oxygen) derived from any of the above feedstock streams. Transforming these "waste streams" into energy can help create additional revenue for rural economies and solve waste-disposal problems.

Solid Biofuel Production

Lifeasible offers quality technical consulting, project design, and process optimization services for two kinds of solid biofuel production technologies—biomass solidification technology and pyrolysis technology. These soild biofuel production technologies can convert biomass materials to soild biofuels to make them more convenient to use.

With the rapid consumption of coal energy worldwide, existing solid non-renewable energy sources are gradually depleted. Faced with this situation, there is an urgent need to develop solid renewable energy to reduce the dependence on traditional solid fuel and mitigate the impact of traditional solid fuel use on the environment. Solid biofuel is a renewable, environment-friendly clean energy and a major component of renewable energy that with industrialization and large-scale production prospects. Therefore, Lifeasible is committed to helping customers improve the level of technologies in solid biofuel production, and helping them to accelerate technological innovation and access high quality products. Solid biofuel, compared to liquid biofuel or gaseous biofuel, is often cheaper, easier to extract, and more stable to transport. Generally, wood, branches, straw and sawdust also belong to solid biofuel, it is convenient to use them as fuel, but it is not convenient to transfort due to different shapes. Furthermore, the environmental burden may be increased when burning them directly. Therefore, in order to obtain environmentally friendly solid biofuel, these biomass materials need to be processed. The methods for solid biofuel production usually include biomass solidification technology and pyrolysis technology [6].

Solid biofuels covers solid organic, non-fossil material of biological origin (also known as biomass) which may be used as fuel for heat production or electricity generation. In energy statistics, solid biofuels is a product aggregate equal to the sum of charcoal, fuelwood, wood residues and by-products, black liquor, bagasse, animal waste, other vegetal materials and residuals and renewable fraction of industrial waste [7].

Heat Transfer and Electrical Power

Thermoelectric devices are made from materials that can convert a temperature difference into electricity, without requiring any moving parts — a quality that makes thermoelectrics a potentially appealing source of electricity. The phenomenon is reversible: If electricity is applied to a thermoelectric device, it can produce a temperature difference. Today, thermoelectric devices are used for relatively low-power applications, such as powering small sensors along oil pipelines, backing up batteries on space probes, and cooling minifridges. Any matter which is made up of atoms and molecules has the ability to transfer heat. The atoms are in different types of motion at any time. The motion of molecules and atoms is responsible for heat or thermal energy and every matter has this thermal energy. The more the motion of molecules, more will be the heat energy. However, talking about heat transfer, it is nothing but the process of transfer of heat from a high-temperature body to a low temperature one.

How is Heat Transferred?
Heat can travel from one place to another in several ways. The different modes of heat transfer include:

- Conduction
- Convection
- Radiation

Meanwhile, if the temperature difference exists between the two systems, heat will find a way to transfer from the higher to the lower system.

What Is Conduction?
Conduction is defined as:

The process of transmission of energy from one particle of the medium to another with the particles being in direct contact with each other.

An area of higher kinetic energy transfers thermal energy towards the lower kinetic energy area. High-speed particles clash with particles moving at a slow speed, as a result, slow speed particles increase their kinetic energy. This is a typical form of heat transfer and takes place through physical contact. Conduction is also known as thermal conduction or heat conduction.

Conduction Examples
Following are the examples of conduction:

- Ironing of clothes is an example of conduction where the heat is conducted from the iron to the clothes.
- Heat is transferred from hands to ice cube resulting in the melting of an ice cube when held in hands.
- Heat conduction through the sand at the beaches. This can be experienced during summers. Sand is a good conductor of heat.

What is Convection?
Convection is defined as

The movement of fluid molecules from higher temperature regions to lower temperature regions.

Convection Examples

Examples of convection include:

- Boiling of water, that is molecules that are denser move at the bottom while the molecules which are less dense move upwards resulting in the circular motion of the molecules so that water gets heated.
- Warm water around the equator moves towards the poles while cooler water at the poles moves towards the equator.
- Blood circulation in warm-blooded animals takes place with the help of convection, thereby regulating the body temperature.

What Is Radiation?

Radiant heat is present in some or other form in our daily lives. Thermal radiations are referred to as radiant heat. Thermal radiation is generated by the emission of electromagnetic waves. These waves carry away the energy from the emitting body.

Radiation takes place through a vacuum or transparent medium which can be either solid or liquid. Thermal radiation is the result of the random motion of molecules in matter. The movement of charged electrons and protons is responsible for the emission of electromagnetic radiation. Let us know more about radiation heat transfer.

Radiation heat transfer is measured by a device known as thermocouple. A thermocouple is used for measuring the temperature. In this device sometimes, error takes place while measuring the temperature through radiation heat transfer.

Radiation Example:

Following are the examples of radiation:

- Microwave radiation emitted in the oven is an example of radiation.
- UV rays coming from the sun is an example of radiation.
- The release of alpha particles during the decaying of Uranium-238 into Thorium-234 is an example of radiation [8].

Electric power:

Electric power is the rate at which electrical energy is transferred by an electric circuit. The SI unit of power is the watt, one joule per second. Standard suffixes apply to watts as with other SI units: thousands, millions and billions of watts are called kilowatts, megawatts and gigawatts respectively.

An important aspect of any electrical or electronic circuit is the power associated with it. It is found that when a current flows through a resistor, electrical energy is converted into heat. This fact is used by electrical heaters which consist of a resistor through which current flows. Light bulbs use the same principle, heating the element up so that it glows white hot and produces light. At other times much smaller resistors and very much smaller currents are used. Here the amount of heat generated may be very small. However if some current flows then some heat is generated. In this instance the heat generated represents the amount of electrical power being dissipated.

Definition of power:
Whether power is used in a mechanical environment or an electrical environment, the definition of power is still the same. The way in which it may be discussed may be slightly different, but nevertheless the definition and actuality of it is exactly the same.

Calculating power:
The amount of power dissipated in a circuit can be easily determined. It is simply the product of the potential difference or voltage across the particular element, multiplied by the current flowing through it. In other words an electrical fire running from a 250 volt supply, and consuming 4 amps of current will dissipate 250 x 4 = 1000 watts or 1 kilowatt. In other words [9].

$$W=V\ I$$

Solar Thermal Energy:
Basically, solar energy is produced by capturing radiant light and heat from the sun and then converting it into energy to power homes, businesses, tools and machinery. Electricity is generated during peak periods — hot afternoons — when the demand is highest and air conditioners are going full blast.

Thermal energy is one of the most important energy sources for daily usage, industrial processes heat, and to generate electricity. Global energy demand continues to increase with the expansion of the economy activities and sustainable development where the world is suffering from pollution and harmful greenhouse gases due to over burning of fossil fuels. There are different form of energies and ways to convert one form to other form. Recently, conventional thermal energy is integrated with solar energy to the heat demand and to produce electricity. Combined heat and power are another efficient use of thermal energy. Currently, the researchers are more concentrated to analyze the performance of thermal energy conversion systems all over the world. Solar thermal energy (STE) is a form of energy and a technology for harnessing solar energy to generate thermal energy for use in industry, and in the residential and commercial sectors.

Solar thermal collectors are classified by the United States Energy Information Administration as low-, medium-, or high-temperature collectors. Low-temperature collectors are generally unglazed and used to heat swimming pools or to heat ventilation air. Medium-temperature collectors are also usually flat plates but are used for heating water or air for residential and commercial use.

High-temperature collectors concentrate sunlight using mirrors or lenses and are generally used for fulfilling heat requirements up to 300 deg C / 20 bar pressure in industries, and for electric power production. Two categories include Concentrated Solar Thermal (CST) for fulfilling heat requirements in industries, and Concentrated Solar Power (CSP) when the heat collected is used for electric power generation. CST and CSP are not replaceable in terms of application [10].

Solar Photovoltaic Energy Conversion:
Photovoltaic cells convert sunlight into electricity

A photovoltaic (PV) cell, commonly called a solar cell, is a nonmechanical device that converts sunlight directly into electricity. Some PV cells can convert artificial light into electricity.

1.11 What is Solar PV?

Solar Photovoltaic (PV) is a technology that converts sunlight (solar radiation) into direct current electricity by using semiconductors. When the sun hits the semiconductor within the PV cell, electrons are freed and form an electric current.

Solar PV technology is generally employed on a panel (hence solar panels). PV cells are typically found connected to each other and mounted on a frame called a module. Multiple modules can be wired together to form an array, which can be scaled up or down to produce the amount of power needed.

PV cells can be made from various semi-conductor materials. The most commonly used material today is silicon but other materials, such the ones listed below, are being tested and used to increase the efficiency of converting sunlight to electricity.

- Monocrystalline Silicon
- Polycrystalline Silicon
- Amorphous Silicon
- Cadmium Telluride (CdTe)
- Copper Indium Gallium Selenide (CIGS)

Almost 90% of the world's PV technologies, today, are based on some variation of silicon1. In 2011, about 95% of all shipments by U.S. manufacturers to the residential sector were crystalline silicon solar panels [11].

1.12 Solar Photovoltaic Energy Conversion

In spite of high specific costs, photovoltaic systems provide an increasingly attractive alternative for electricity supply in particular at remote locations, due to their extremely high reliability, low maintenance requirements, long lifetime and their modularity with flexible system sizing down to very small load demands.

The special attraction of photovoltaics, as compared to other power generation technologies, lies in the fact that the solar radiation is converted directly into electric power by an electronic solid state process. In general, no moving parts and no specific thermal stresses are involved. Therefore, photovoltaic systems operate quietly and they can offer extremely high reliability, low maintenance requirements and a long lifetime. Due to the nature of the conversion process, one can utilize direct as well as diffuse radiation, which also allows applications in moderate climates with higher fractions of diffuse radiation. Another important advantage of PV is its modularity, permitting a very flexible system sizing for integration into buildings and for decentral applications down to very small load demands [12].

Electrochemical Energy Conversion and Storage:
Electrochemical energy conversion is a field of energy technology concerned with electrochemical methods of energy conversion including fuel cells and photoelectrochemical. This field of technology also includes electrical storage devices like batteries and supercapacitors. Electrochemical cells and systems play a key role in a wide range of industry sectors. These devices are critical enabling technologies for renewable energy; energy management, conservation, and storage; pollution control/monitoring;

and greenhouse gas reduction. A large number of electrochemical energy technologies have been developed in the past. These systems continue to be optimized in terms of cost, life time, and performance, leading to their continued expansion into existing and emerging market sectors. The more established technologies such as deep-cycle batteries and sensors are being joined by emerging technologies such as fuel cells, large format lithium-ion batteries, electrochemical reactors; ion transport membranes and supercapacitors. This growing demand (multi billion dollars) for electrochemical energy systems along with the increasing maturity of a number of technologies is having a significant effect on the global research and development effort which is increasing in both in size and depth. A number of new technologies, which will have substantial impact on the environment and the way we produce and utilize energy, are under development. This paper presents an overview of several emerging electrochemical energy technologies along with a discussion some of the key technical challenges.

1.13 What is Waste to Energy?

Waste to Energy (WTE), is a term that is used to describe various technologies that convert non-recyclable waste into usable forms of energy including heat, fuels and electricity. WTE can occur through a number of processes such as incineration, gasification, pyrolysis, anaerobic digestion, and landfill gas recovery.

The term WTE is commonly used in specific reference to incineration which burns completely combusted waste at ultra-high temperatures allowing for energy recovery. Modern incineration facilities use pollution control equipment to prevent the release of emissions into the environment. Currently incineration is the only WTE technology that is economically viable and operationally feasible at commercial scale.

Another example of WTE is anaerobic digestion (AD), an old but effective technology that biologically converts organic material into compost as well as biogas for energy. AD systems have large potential and can range from low to high tech, therefore they can service communities of all income levels. Another process, called pyrolysis, can thermo-chemically convert waste products into clean liquid fuels.

Finally, landfill gas recovery refers to the process of capturing the gases emitted from municipal landfills and converting it for energy. The most common form of collection occurs by drilling horizontal or vertical wells into the landfill and uses blowers and vacuums to collect the gas for treatment [13].

Every year, millions of tons of solid waste is generated in households around the world and which is disposed of in council and municipal landfill sites, while at the same time, millions of units of energy from coal, oil and natural gas are burned daily to create the electrical power we need to run our homes and workplaces. Today, a new generation of Waste to Energy, (WTE) conversion technologies are emerging which hold the potential to create clean renewable energy from solid and liquid waste materials.

Waste to energy—The energy content available from combustion of solid waste represents a significant "alternative energy" supply to help reduce the use and our dependency on conventional fossil fuels. "Waste" by definition includes any waste materials other than liquids, gases and any non-usable by-product generated as a result of a process or production that is no longer deemed valuable, and are therefore normally discarded.

Such waste materials typically originate from either the residential community as municipal solid waste or from commercial, light-industrial, and agricultural wastes. Wastes generated from heavy industrial manufacturing activities, chemical and medical industries are typically classified as hazardous wastes and are generally not used to convert this type of waste to energy. Traditionally the term "waste to energy" has generally referred to the practice of incineration of waste products either by burning the rubbish in the back yard or by large industrial incinerators to produce heat. The category of waste to energy broadly describes any of a number of processes or technologies in which a useful byproduct (energy) is recovered from an otherwise unusable source.

Waste to energy technologies—Physically convert waste matter into more useful forms of fuel that can be used to supply energy. These waste to energy processes includes thermal conversions such as combustion (incineration), pyrolysis, gasification, or biological treatments such as anaerobic digestion and fermentation, etc, and various combinations of the above.

For example, solid wastes can be converted into fuel pellets and along with gasified waste co-fired with fossil fuel coal in an existing conventional coal fired power station. Energy can be derived from waste in a number of ways. Waste to energy processes include wastes that has been treated and made into a solid fuel for incineration to produce heat and steam. Waste that has been converted into biogas from both organic and inorganic wastes, or biological technologies, in which bacterial fermentation is used to digest organic wastes to yield fuel, converting the waste to fuel [14].

Polymer and Composite Materials for Renewable Energy:
Many high performance composites exist, yet most notable for renewable energy structures are variations of fiber-reinforced polymers (FRPs), carbon-fiber-reinforced polymers (CFRPs) and glass-reinforced plastic (GRPs or e-glass) are leaders in the ever advancing materials engineering sectors. Renewable energy is at the forefront of the energy policy of all major economies. It not only helps to reduce the dependence on fossil fuel but also helps to reduce the negative impact on the environment and economy.

A major portion of greenhouse gas is carbon dioxide, pumped into the environment by industries, specifically the power sector. Major sources of carbon dioxide gas are the burning of fossil fuels in the production of electricity and heat in various industries [15].

Polymer Materials in Solar Energy Industry:
As per IEA, 114.9 GW of new solar power installations happened globally in 2019. The year 2019 shows a 12% growth over 2018 in terms of new installations. The solar energy constitutes around 3% of global electricity demand in 2019.

According to IEA, the new solar installations are expected to decline by 12% in 2020 as compared to 2019 installations, due to the Covid-19 pandemic. The solar installations are estimated to rebound as majority of the projects in pipeline are financed and can start construction as the pandemic situation gets normal.

As PV, Solar Thermoelectric Generators (STEG), PV/T, and concentrated or conventional PV systems integrated with STEG, STC, and energy storage can lead to an increase in the electrical and thermal energy generated and in system lifetime. Hence PV-based configurations and hybrid systems demand increased.

Due to the new advancements, solar technology is set to become lighter, more flexible, and applicable everywhere. These advances include:

- Floating solar farms
- BIPV solar technology
- Solar skins
- Solar fabric, and
- Photovoltaic solar noise barriers (PVNB) [15].

Algal Energy Technology:
Algae fuel, algal biofuel, or algal oil is an alternative to liquid fossil fuels that uses algae as its source of energy-rich oils. Also, algae fuels are an

alternative to commonly known biofuel sources, such as corn and sugarcane. When made from seaweed (macroalgae) it can be known as seaweed fuel or seaweed oil. Energy is the utmost requirement for driving the organization and maintenance of entire ecosystem. Our continued dependency on fossil fuels such as coal, petroleum and natural gas as the prime source of energy has led to serious concerns about the future energy supply and security. Furthermore, over-consumption of carbon-based fossil energy sources raises serious environmental issues of global warming and climate change. To overcome the global energy demand and to enable economic as well as ecological development in a sustainable manner, technological progress for the utilization of renewable natural energy are essential to protect the environment and save energy in today's increasingly competitive world. To this end, algal biofuels are being claimed as an apt alternative energy source and in recent past, several taxonomic groups of algae have been studied and reported as an alternative to fossil fuels. It is envisaged that algal biomass could be readily processed into the raw material to make cost-effective biofuels and is being explored as an emergent and renewable green energy crops for the production of biofuels, especially biodiesel. Development of astonishing technological innovations in the field of algal genetic engineering has triggered remarkable output across the global energy sector for better biofuels. Several new techniques are being adopted for large-scale farming of microalgae intended for biofuel production. However, there are certain constraints for commercial-scale energy production from algae. The present review discusses the technological development and current information on the cultivation and process of biodiesel production form algae. Also, discussed are the technological development and genomic insights into the algal biomass and triacylglycerol accumulation for enhanced biodiesel production.

Algae biofuel: hype and disappointment:

For more than five years from around 2005, algae-based biofuel companies, including the likes of Algenol, Sapphire Energy and Solazyme, raised hundreds of millions of dollars in private sector investment on the promise that chemically engineering algae could scale up to produce tens of millions of gallons of fuel in a matter of years, at prices competitive with fossil fuels. Fuel conversion from algae is broadly based on the feedstock's high concentrations of lipids: fatty, oil-containing acid molecules that can be extracted to create biofuels.

Nearly 15 years later, the green tech world has fallen out of love with algal biofuel. Despite the large sums spent on developing the conversion process, the industry's ambitious production goals—not to mention cost-competitiveness with fossil fuels—remain a distant dream. In terms of cost, major oil price declines in 2008 and 2014 certainly did not help biofuel competitiveness, but technical issues have also proven a major sticking point. Intractable problems have been encountered in terms of the energy balance of lipid extraction, maintaining suitable growing conditions in open ponds, and the immense volumes of water, CO_2 and fertiliser required to allow the algae to photosynthesize fast enough at large scales.

Solving the energy balance issue:
One of the main points of friction in the lipid extraction component of algal biofuel production is the need to remove all moisture from the algae in advance, leaving a dry powder from which the lipids can be separated. This is part of the reason that it often requires more energy to power the process than the resulting fuel provides at the other end of the equation.

However, a new method invented by researchers at the University of Utah may have a solution to the energy parity conundrum. A team of chemical engineers have developed a new jet mixer technology that precludes the need for energy-intensive drying processes. The new mixing reactor shoots jets of solvent into jets of algae in liquid suspension, creating the turbulence needed to prompt the lipids to transfer into the solvent stream. Not only does this process require much less energy, the researchers say it is also faster, with lipids extracted in mere seconds [16].

Solar Photovoltaic Power Systems:
Solar PV systems are power systems that convert sunlight into electricity by utilizing the photovoltaic effect. This is a process in which semiconducting materials generate voltage and current when exposed to light. PV modules and arrays are just one part of a PV system. Systems also include mounting structures that point panels toward the sun, along with the components that take the direct-current (DC) electricity produced by modules and convert it to the alternating-current (AC) electricity used to power all of the appliances in your home.

The largest PV systems in the country are located in California and produce power for utilities to distribute to their customers. The Solar Star PV power station produces 579 megawatts of electricity, while the Topaz Solar Farm and Desert Sunlight Solar Farm each produce 550 megawatts.

How Solar PV Panels Generate Electricity?
Solar photovoltaic (PV) panels use cells containing a semi-conductor material to capture the sun's energy and convert solar radiation into electricity.

The most commonly used semi-conductor material is silicon, which is an abundant natural resource found in sand. When light strikes the cell a certain amount of energy is absorbed within the semiconductor material, knocking electrons, the negatively charged particles that form the basis of electricity, loose.

Most PV cells have two layers of semi-conductor material, one positively charged and one negatively charged. When light shines on the semi-conductor the electric field across the junction between these two layers causes electricity to flow, generating direct current (DC). By placing metal contacts on the top and bottom of the PV cell, we can draw that current off for external use.

Solar PV electric panels do not require bright sunlight in order to operate, meaning that you can still generate electricity on cloudy days, however in general the greater the intensity of light the higher the flow of electricity. Although, due to the reflection of sunlight, days with slight cloud can result in higher energy yields than days with a completely cloudless sky.

It is important to realize that you can only use your free solar electricity when it is being generated—so unless you also invest in batteries to store power for use in the evenings and at night, you will need to pay for your energy use as normal when the panels are not producing electric [17].

How the Electricity is Converted for use?

Direct Current (DC) needs to be converted into alternating current (AC) so it can be used in a domestic building; this is performed by an inverter. The AC electricity then passes via the generation meter, which measures how much electricity has been created, and on to the consumer unit where it can be fed into the property for use or exported back to the national grid via the electricity meter.

Functioning of the photovoltaic cells:

The word "photovoltaic" consists of two words: photo, a greek word for light, and voltaic, which defines the measurement value by which the activity of the electric field is expressed, i.e. the difference of potentials. Photovoltaic systems use cells to convert sunlight into electricity. Converting solar energy into electricity in a photovoltaic installation is the most known way of using solar energy. The light has a dual character according to quantum physics. Light is a particle and it is a wave. The particles of light are called photons. Photons are massless particles, moving at light speed. The energy of the photon depends on its wavelength and the frequency, and we can calculate it by the Einstein's law, which is: where: E - photon energy h - Planck's constant = 6.626×10 – 34Js - photon frequency. In metals and in the matter generally, electrons can exist as valence or as free. Valence electrons are associated with the atom, while the free electrons can move freely. In order for the valence electron to become free, he must get the energy that is greater than or equal to the energy of binding. Binding energy is the energy by which an electron is bound to an atom in one of the atomic bonds. In the case of photoelectric effect, the electron acquires the required energy by the collision with a photon. Part of the photon energy is consumed for the electron getting free from the influence of the atom which it is attached to, and the remaining energy is converted into kinetic energy of a now free electron. Free electrons obtained by the photoelectric effect are also called photoelectrons. The energy required to release a valence electron from the impact of an atom is called a "work out", Wi, and it depends on the type of material in which the photoelectric effect has occurred [18].

Wind Energy and Small Hydro Power Systems:
Uses of Hydropower In the United States today, hydropower projects provide 81% of the nation's renewable electricity generation and about 10% of the nation's total electricity. That is enough to power 37.8 million homes, according to the National Hydropower Association. The vast majority of the hydropower produced in the United States comes from large-scale projects that generate more than 30 megawatts (MW)—enough electricity to power nearly 30,000 households. Small-scale hydropower systems are those that generate between .01 and 30 MW of electricity. Hydropower systems that generate up to 100 kilowatts (kW) of electricity are often called microhydro systems. Most of the systems used by home and small business owners would qualify as microhydro systems. In fact, a 10 kW system generally can provide enough power for a large home, a small resort, or a hobby farm.

How Hydropower Works?
Hydropower systems use the energy in flowing water to produce electricity or mechanical energy. Although there are several ways to harness the moving water to produce energy, run-of-the-river systems, which do not require large storage reservoirs, are often used for microhydro, and sometimes for small-scale hydro, projects. For run-of-the-river hydro projects, a portion of a river's water is diverted to a channel, pipeline, or pressurized pipeline (penstock) that delivers it to a waterwheel or turbine. The moving water rotates the wheel or turbine, which spins a shaft. The motion of the shaft can be used for mechanical processes, such as pumping water, or it can be used to power an alternator or generator to generate electricity [19].

Is Hydropower Right for You? Of course to build a small hydropower system, you need access to flowing water. A sufficient quantity of falling water must be available, which usually, but not always, means that hilly or mountainous sites are best. Next you will want to determine the amount of power that you can obtain from the flowing water on your site. The power available at any instant is the product of what is called flow volume and what is called head.

System Components Small run-of-the-river hydropower systems consist of these basic components:

- Water conveyance—channel, pipeline, or pressurized pipeline (penstock) that delivers the water
- Turbine or waterwheel—transforms the energy of flowing water into rotational energy
- Alternator or generator—transforms the rotational energy into electricity
- Regulator—controls the generator
- Wiring—delivers the electricity

A Clean Energy Future By investing in a small hydropower system, you can reduce your exposure to future fuel shortages and price increases, and help reduce air pollution. There are many factors to consider when buying a system, but with the right site and equipment, careful planning, and attention to regulatory and permit requirements, small hydropower systems can provide you a clean, reliable source of power for years to come [19].

Advanced Battery and Fuel Cell Technologies:
Secondary cells:
A secondary cell, or storage battery, can function as either a voltaic or an electrolytic cell. When its electrodes are connected to a device that uses electricity, it functions like any other battery. When a current is fed through it, the electrochemical reactions are reversed, replenishing the battery's strength.

Lead-acid storage batteries: Getting started:
In 1866 the Frenchman Gaston Planté discovered the combination still widely used today for storage batteries: electrodes of rough lead and lead dioxide immersed in a solution of sulfuric acid.

With the coming of large generators in the 1880s, storage batteries became tools for industrial and consumer use. In power systems batteries were charged by generators during the daylight hours and discharged at night when people turned on their lights. In trolley systems they equalized the uneven demand for power caused by starting and stopping in traffic. Lead-acid storage batteries began to appear nearly everywhere.

The most familiar use today for lead-acid storage batteries is to start car engines. Early automobiles were started by hand-cranking the engine. When Charles F. Kettering invented the electric starter in 1911, his boss — who had recently broken his arm while starting his car — was eager to finance this new invention. Kettering's device was powered by a rechargeable lead-acid battery. Demand for batteries shot up overnight: Kettering's first order for ten thousand batteries from the Electric Storage Battery Company of Philadelphia was greeted in disbelief.

Advanced battery systems:
For large-scale energy applications to replace the lead-acid storage battery, scientists have turned again to lithium and sodium for their high reactivity and their relative abundance. The Electrolytes needed for large-scale lithium and sodium batteries only work at high temperatures, however. Developing other high-energy battery systems remains an objective for many research and development teams.

Fuel cells: An alternative to combustion:
As far back as 1839, the English scientist William Grove had the idea that the reactants of a battery could be gases fed into it from external tanks, For most of their history, fuel cells existed only as laboratory curiosities. Then in the 1950s in the United States, Francis T. Bacon developed the first operational multikilowatt cell, with an output of five kilowatts. The first major application of fuel cells was in the Gemini and Apollo space flights in the 1960s.

Fuel cells have also been considered as power sources for vehicles and for remote and back-up power stations. In the fuel cell's basic chemical reactions, the electrons transferred from atoms of fuel (e.g. hydrogen) to oxygen atoms are immediately available as electrical energy through the external circuit that connects the cell's electrodes. The process is much cleaner and more efficient than igniting fuel in an internal combustion engine or burning it to heat water for a steam engine and using that energy to turn electric generators [20].

Nanotechnology for Energy Systems:
Nanotechnology is being used to reduce the cost of catalysts used in fuel cells. These catalysts produce hydrogen ions from fuel such as methanol.

Nanotechnology is also being used to improve the efficiency of membranes used in fuel cells to separate hydrogen ions from other gases, such as oxygen. Nanotechnologies provide the potential to enhance energy efficiency across all branches of industry and to economically leverage renewable energy production through new technological solutions and optimized production technologies.

Energy sources:
Nanotechnologies provide essential improvement potentials for the development of both conventional energy sources (fossil and nuclear fuels) and renewable energy sources like geothermal energy, sun, wind, water, tides or biomass. Nano-coated, wear resistant drill probes, for example, allow the optimization of lifespan and efficiency of systems for the development of oil and natural gas deposits or geothermal energy and thus the saving of costs. Further examples are high-duty nanomaterials for lighter and more rugged rotor blades of wind and tide power plants as well as wear and corrosion protection layers for mechanically stressed components (bearings, gear boxes, etc.). Nanotechnologies will play a decisive role in particular in the intensified use of solar energy through photovoltaic systems. In case of conventional crystalline silicon solar cells, for instance, increases in efficiency are achievable by anti reflection layers for higher light yield.

First and foremost, however, it will be the further development of alternative cell types, such as thin-layer solar cells (among others of silicon or other material systems like copper/indium/selenium), dye solar cells or polymer solar cells, which will predominantly profit from nanotechnologies. Polymer solar cells are said to have high potential especially regarding the supply of portable electronic devices, due to the reasonably-priced materials and production methods as well as the flexible design. Medium-term development targets are an efficiency of approx. 10% and a lifespan of several years. Here, for example, nanotechnologies could contribute to the optimization of the layer design and the morphology of organic semiconductor mixtures in component structures. In the long run, the utilization of nanostructures, like quantum dots and wires, could allow for solar cell efficiencies of over 60%.

Energy conversion:
The conversion of primary energy sources into electricity, heat and kinetic energy requires utmost efficiency. Efficiency increases, especially in fossil-fired gas and steam power plants, could help avoid considerable amounts of carbon dioxide emissions.

Higher power plant efficiencies, however, require higher operating temperatures and thus heat-resistant turbine materials. Improvements are possible, for example, through nano-scale heat and corrosion protection layers for turbine blades in power plants or aircraft engines to enhance the efficiency through increased operating temperatures or the application of lightweight construction materials (e.g. titanium aluminides).

Nano-optimized membranes can extend the scope of possibilities for separation and climate-neutral storage of carbon dioxide for power generation in coal-fired power plants, in order to render this important method of power generation environmentally friendlier in the long run. The energy yield from the conversion of chemical energy through fuel cells can be stepped up by nano-structured electrodes, catalysts and membranes, which results in economic application possibilities in automobiles, buildings and the operation of mobile electronics.

Energy distribution

Regarding the reduction of energy losses in current transmission, hope exists that the extraordinary electric conductivity of nanomaterials like carbon nanotubes can be utilized for application in electric cables and power lines. Furthermore, there are nanotechnological approaches for the optimization of superconductive materials for lossless current conduction.

In the long run, options are given for wireless energy transport, e.g. through laser, microwaves or electromagnetic resonance. Future power distribution will require power systems providing dynamic load and failure management, demand-driven energy supply with flexible price mechanisms as well as the possibility of feeding through a number of decentralized renewable energy sources.

Nanotechnologies could contribute decisively to the realization of this vision, inter alia, through nano-sensory devices and power-electronical components able to cope with the extremely complex control and monitoring of such grids.

Energy storage:

The utilization of nanotechnologies for the enhancement of electrical energy stores like batteries and super-capacitors turns out to be downright promising. Due to the high cell voltage and the outstanding energy and power density, the lithium-ion technology is regarded as the most promising variant of electrical energy storage.

Nanotechnologies can improve capacity and safety of lithium-ion batteries decisively, as for example through new ceramic, heat-resistant and still flexible separators and high-performance electrode materials. The company Evonik pushes the commercialization of such systems for the application in hybrid and electric vehicles as well as for stationary energy storage.

In the long run, even hydrogen seems to be a promising energy store for environmentally-friendly energy supply. Apart from necessary nanostructure adjustments, the efficient storage of hydrogen is regarded as one of the critical factors of success on the way to a possible hydrogen management.

Current materials for chemical hydrogen storage do not meet the demands of the automotive industry, which requires a hydrogen-storage capacity of up to ten weight percent.

Energy usage:

To achieve sustainable energy supply, and parallel to the optimized development of available energy sources, it is necessary to improve the efficiency of energy use and to avoid unnecessary energy consumption. This applies to all branches of industry and private households. Nanotechnologies provide a multitude of approaches to energy saving.

Examples are the reduction of fuel consumption in automobiles through lightweight construction materials on the basis of nanocomposites, the optimization in fuel combustion through wear-resistant, lighter engine components and nanoparticular fuel additives or even nanoparticles for optimized tires with low rolling resistance.

Considerable energy savings are realizable through tribological layers for mechanical components in plants and machines. Building technology also provides great potentials for energy savings, which could be tapped, for example, by nanoporous thermal insulation material suitably applicable in the energetic rehabilitation of old buildings.

Nanomaterials:

Scientists have not unanimously settled on a precise definition of nanomaterials, but agree that they are partially characterized by their tiny size, measured in nanometers. A nanometer is one millionth of a millimeter—approximately 100,000 times smaller than the diameter of a human hair.

Nano-sized particles exist in nature and can be created from a variety of products, such as carbon or minerals like silver, but nanomaterials by definition must have at least one dimension that is less than approximately 100 nanometers. Most nanoscale materials are too small to be seen with the naked eye and even with conventional lab microscopes.

Materials engineered to such a small scale are often referred to as engineered nanomaterials (ENMs), which can take on unique optical, magnetic, electrical, and other properties. These emergent properties have the potential for great impacts in electronics, medicine, and other fields. For example,

1. Nanotechnology can be used to design pharmaceuticals that can target specific organs or cells in the body such as cancer cells, and enhance the effectiveness of therapy.
2. Nanomaterials can also be added to cement, cloth and other materials to make them stronger and yet lighter.
3. Their size makes them extremely useful in electronics, and they can also be used in environmental remediation or clean-up to bind with and neutralize toxins.

1.14 Where are Nanomaterials Found?

Some nanomaterials can occur naturally, such as blood borne proteins essential for life and lipids found in the blood and body fat. Scientists, however, are particularly interested in engineered nanomaterials (ENMs), which are designed for use in many commercial materials, devices and structures. Already, thousands of common products—including sunscreens, cosmetics, sporting goods, stain-resistant clothing, tires, and electronics—are manufactured using ENMs. They are also in medical diagnosis, imaging and drug delivery and in environmental remediation.

Advantages of Nanomaterials:
The properties of nanomaterials, particularly their size, offer various different advantages compared to the bulk-form of the materials, and their versatility in terms of the ability to tailor them for specific requirements accentuates their usefulness. An additional advantage is their high porosity, which again increases demand for their use in a multitude of industries.

In the energy sector, the use of nanomaterials is advantageous in that they can make the existing methods of generating energy—such as solar panels—more efficient and cost-effective, as well as opening up new ways in which to both harness and store energy.

Nanomaterials are also set to introduce a number of advantages in the electronics and computing industry. Their use will permit an increase in the accuracy of the construction of electronic circuits on an atomic level, assisting in the development of numerous electronic products.

The very large surface-to-volume ratio of nanomaterials is especially useful in their use in the medical field, which permits the bonding of cells and active ingredients. This results in the obvious advantage of an increase in the likelihood of successfully combatting various diseases.

Disadvantages of Nanomaterials:

Alongside their benefits, there are also a number of disadvantages associated with nanomaterial use. Due to the relative novelty of the widespread use of nanomaterials, there is not a large amount of information on the health and safety aspects of exposure to the materials.

Currently, one of the main disadvantages associated with nanomaterials is considered to be inhalation exposure. This concern arises from animal studies, the results of which suggested that nanomaterials such as carbon nanotubes and nanofibers may cause detrimental pulmonary effects, such as pulmonary fibrosis. Further possible health risks are ingestion exposure and dust explosion hazards.

Additionally, there are still knowledge gaps regarding nanomaterials, meaning the manufacturing process can often be complex and difficult. The overall process is also expensive, requiring optimum results—especially regarding their use in consumer goods—in order to avoid financial losses.

Risk-assessments concerning any potential environmental effects indicate that nanomaterials used in cosmetic items such as sunscreen, which are applied to the skin, run the risk of ending up in aquatic ecosystems after they are washed off. Nanomaterials that have been engineered may also end up in water bodies such as lakes and rivers, before accumulating to create particles of a larger size. This may put freshwater species—such as snails—at risk by possibly inducing a decline in life processes such as growth and reproduction. The same issues caused by the materials in such freshwater ecosystems are likely to pertain to marine ecosystems as well. Accumulation of nanomaterials in other aspects of the environment, such as soils—through sewage sludge—is an additional concern. Although the concentrations of these engineered nanomaterials is expected to be quite small, repeated release may cause the concentrations to increase over time, exacerbating the related negative effects [21].

Properties of nanomaterials:

As the size is reduced to the nanometric scale, the exposed surface area increases and this favors the greater interaction between nearby atoms and molecules, giving rise to various interactions, attractions and repulsions that cause surface, electronic and quantum effects that affect to the optical, electric and magnetic behaviors of the materials.

This means that with a very small amount of nanomaterial the properties of other materials can be modified and significantly improved, providing great potential and added value. An example of this will be polymers doped with carbon nanotubes, which make the doped material have a lightness, mechanical strength and functionality superior to metals.

Processing of green energy materials

Sustainable Materials and Green Processing for Energy Conversion provides a concise reference on green processing and synthesis of materials required for the next generation of devices used in renewable energy conversion and storage. The book covers the processing of bio-organic materials, environmentally-friendly organic and inorganic sources of materials, synthetic green chemistry, bioresorbable and transient properties of functional materials, and the concept of sustainable material design. The book features chapters by worldwide experts and is an important reference for students, researchers, and engineers interested in gaining extensive knowledge concerning green processing of sustainable, green functional materials for next generation energy devices. Additionally, functional materials used in energy devices must also be able to degrade and decompose with minimum energy after being disposed of at their end-of-life. Environmental pollution is one of the global crises that endangers the life cycles of living things. There are multiple root causes of this pollution, including industrialization that demands a huge supply of raw materials for the production of products related to meeting the demands of the Internet-of-Things. As a result, improvement of material and product life cycles by incorporation of green, sustainable principles is essential to address this challenging issue.

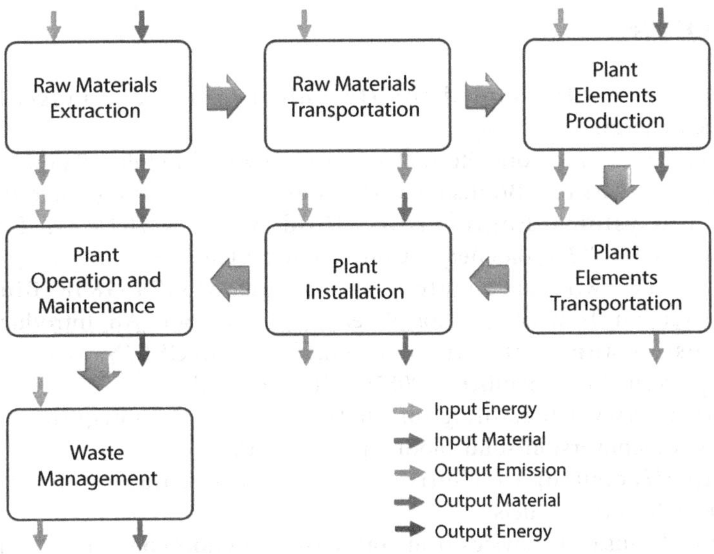

Green energy is at the heart of all ecological strategies because it affects companies in three vital areas: environmental, economic, and social. Conventional energy sources based on oil, coal, and natural gas have proven to be highly effective drivers of economic progress, but at the same time damaging to the environment and to human health. The potential of renewable energy sources is enormous as they can in principle meet many times the world's energy demand. Renewable energy sources, such as biomass, wind, solar, hydropower, and geothermal can provide sustainable energy services, based on the use of routinely available, indigenous resources. Renewable energy sources currently supply somewhere between 15% and 20% of world's total energy demand. The supply is dominated by traditional biomass, mostly fuel wood used for cooking and heating, especially in developing countries in Africa, Asia and Latin America. A major contribution is also obtained from the use of large hydropower; with nearly 20% of the global electricity supply being provided by this source. New renewable energy sources (solar energy, wind energy, modern bio-energy, geothermal energy, and small hydropower) are currently contributing about 2%. A number of scenario studies have investigated the potential contribution of renewables to global energy supplies, indicating that in the second half of the 21st century their contribution might range from the present figure of nearly 20% to more than 50% with the right policies in place.

References

1. https://www.twi-global.com/technical-knowledge/faqs/what-is-green-energy.
2. https://www.unido.org/sites/default/files/2009-02/Module7_0.pdf.
3. eng.libretexts.org/Bookshelves/Biological_Engineering/Introduction_to_Biosystems_Engineering_(Holden_et_al.)/01%3A_Energy_Systems/1.01%3A_Bioenergy_Conversion_Systems.
4. https://www.researchgate.net/publication/347459848_Biomass_Conversion_Technologies_for_Bioenergy_Generation_An_Introduction.
5. https://stannescet.ac.in/cms/staff/qbank/CSE/Notes/CY8151-Engineering%20Chemistry-1908708516-unit_4.pdf.
6. https://www.lifeasible.com/custom-solutions/energy/bioenergy/biomass-conversion/solid-biofuel-production/.
7. https://ec.europa.eu/eurostat/statistics-explained/index.php?title=Glossary:Biofuels.
8. https://byjus.com/physics/heat-transfer-conduction-convection-and-radiation/.
9. https://www.electronics-notes.com/articles/basic_concepts/power/what-is-electrical-power-basics-tutorial.php.
10. https://www.sciencedirect.com/science/article/pii/B978012823 9599000039.

11. https://studentenergy.org/conversion/solar-pv.
12. http://www.desware.net/sample-chapters/d06/d10-014.pdf.
13. https://studentenergy.org/conversion/waste-to-energy.
14. https://madhavuniversity.edu.in/waste-to-energy-conversion.html.
15. https://omnexus.specialchem.com/tech-library/article/growth-of-polymers-and-plastics-used-in-renewable-energy-sector.
16. https://www.power-technology.com/analysis/algal-biofuels-challenges-opportunities/.
17. http://www.ews-solarpower.co.uk/24-how-does-the-system-work.
18. https://www.irena- istra.hr/uploads/media/Photovoltaic_systems.pdf.
19. https://www.nrel.gov/docs/fy01osti/29065.pdf.
20. https://www.electrochem.org/battery-and-fuel-cell.
21. https://www.niehs.nih.gov/health/topics/agents/sya-nano/index.cfm.

2

Recent Advances in Green Energy Materials: A Review

Mukhtiar Singh, Maninder Singh, Jasvinder Singh*, Mandeep Singh and Harjit Singh

School of Mechanical Engineering, Lovely Professional University, Phagwara, Punjab, India

Abstract

The use of green energy materials has become increasingly important in the transition toward a sustainable future. This review paper provides an overview of recent advances in green energy materials, including solar, wind, hydroelectric, geothermal, and biomass materials. The paper discusses the various types of materials used in each of these categories, their properties, and their applications. Additionally, the paper highlights the economic and environmental benefits of using green energy materials. Overall, this paper provides a comprehensive overview of the current state of green energy materials and their potential for future development. Overall, the paper provides a comprehensive overview of recent advances in green energy materials and their potential applications. It highlights the importance of materials science in the development of sustainable energy technologies and provides insights into the future direction of research in this field. The review aims to inspire researchers to explore new materials and technologies to accelerate the transition toward a sustainable energy future.

Keywords: Green energy, sustainable energy, green materials, wind and solar energy, green manufacturing and environmental issues

**Corresponding author*: jasvinder.15854@lpu.co.in

Chandan Deep Singh, Talwinder Singh and Davinder Singh (eds.) Digitization and Manufacturing Performance: An Environmental Perspective, (51–72) © 2025 Scrivener Publishing LLC

2.1 Introduction

The use of fossil fuels as a primary source of energy has led to various environmental issues, such as global warming, air pollution, and water pollution. To mitigate the impact of these issues, the world is shifting toward renewable energy sources such as solar, wind, and hydroelectric power. The development of green energy materials has played a vital role in this shift. This paper provides a review of recent advances in green energy materials. Green energy materials are materials that are environmentally sustainable and can be used in the generation and storage of renewable energy. They are an important aspect of the transition to a more sustainable and low-carbon energy system. Recent advances in green energy materials have led to the development of more efficient and cost-effective technologies for harnessing renewable energy sources, such as solar, wind, geothermal, and biomass. One of the key areas of research in green energy materials is the development of new materials for solar cells. This includes the use of perovskite materials, organic and hybrid photovoltaic materials, and the development of thin-film solar cells. Perovskite solar cells typically consist of a perovskite material as the light-absorbing layer, a charge-transporting layer, and electrodes. The most commonly used perovskite materials are hybrid organic-inorganic lead halide perovskites, such as methylammonium lead iodide (MAPI) and formamidinium lead iodide (FAPI). Other materials, such as all-inorganic perovskites and mixed-cation perovskites, have also been explored. Perovskite solar cells have shown great promise in recent years due to their high efficiency, low cost, and ease of fabrication [1]. Perovskite solar cells have demonstrated remarkable progress in device performance in recent years, with power conversion efficiencies (PCEs) exceeding 25%. However, the high efficiency is often accompanied by poor stability and reproducibility. Several approaches, such as using alternative materials, optimizing device architectures, and passivating defects, have been proposed to improve the stability and reproducibility of perovskite solar cells. The stability of perovskite solar cells is a critical issue that needs to be addressed before their commercialization. Perovskite materials are prone to degradation under heat, moisture, and light, leading to device failure over time. Several approaches, such as encapsulation, interface engineering, and doping, have been proposed to improve the stability of perovskite solar cells [2]. Perovskite solar cells have the potential to revolutionize the solar industry by offering a low-cost and high-efficiency alternative to traditional silicon solar cells. However, the commercialization of perovskite solar cells still faces several challenges, such as their long-term stability and the need for scalable manufacturing processes that can meet the demands of the solar market [3]. Despite these challenges, the potential advantages of perovskite solar cells for commercial applications are significant. For example, perovskite solar cells can be made using low-cost materials and simple fabrication techniques, which can significantly reduce production costs compared to traditional silicon solar cells [4]. Additionally, perovskite solar cells have demonstrated high power conversion efficiencies, with some laboratory prototypes achieving over 25% efficiency, which is comparable to or even higher than silicon solar cell [5]. Furthermore, perovskite solar cells can be manufactured in various forms, such as flexible or transparent, which opens up new possibilities for their integration into a wide range of products, from building materials to portable electronics. However, the commercialization of perovskite solar cells is still in its early stages, and several key challenges need to be addressed before they can be widely adopted. Some of the key challenges facing the commercialization of perovskite solar cells include improving their long-term stability, which is critical for their deployment in real-world conditions, and developing scalable manufacturing processes that can meet the demands of the solar market. Despite these challenges, the potential benefits of perovskite solar cells are significant, and ongoing research efforts are focused on addressing these challenges and unlocking the full potential of this promising technology. Hybrid and organic photovoltaic materials have also shown potential as a low-cost alternative to traditional silicon solar cells. Another area of research in green energy materials is the development of advanced composite materials for wind turbine blades. Advanced composites can reduce the weight of wind turbine blades while maintaining or improving their performance, resulting in more efficient and cost-effective wind energy systems. Polymer/carbon particle thermoelectric composites are a promising emerging green energy material that has gained attention in recent years due to their potential to convert waste heat into useful electrical energy. This technology involves the use of polymer matrices filled with conductive carbon particles to create a thermoelectric composite material. The mechanism behind the working of thermoelectric composites involves the Seebeck effect, which is the conversion of temperature differences into electrical voltage [6]. The heat absorbed by the material creates a flow of electrical charge, which can then be harvested and used to power devices. The use of polymers and carbon particles in these composites is particularly advantageous because they are lightweight, flexible, and have low thermal conductivity. This makes them suitable for use in a variety of applications, including wearable devices, automotive parts, and building insulation. In addition, polymer/carbon particle thermoelectric composites are considered environmentally friendly, as they can be produced using sustainable materials and processes. They also have the potential to reduce energy waste and improve energy efficiency, making them a promising solution for a more sustainable future. Carbon nanotubes (CNTs) are promising materials for thermoelectric applications due to their unique one-dimensional structure and high mobility. One-dimensional nanomaterials, like CNTs, have been found to exhibit better thermoelectric performance than multi-dimensional materials, which makes them a potential candidate for efficient thermoelectric devices. The thermoelectric efficiency of CNTs can be further enhanced by increasing their mobility along the long direction, which can be achieved by optimizing their structure and synthesis process. Additionally, reducing the diameter of CNTs can help to lower their thermal conductivity and improve their thermoelectric performance. Compared to organic polymers, CNTs have significantly higher carrier mobility and thermal conductivity, which makes them more desirable for thermoelectric applications. However, optimizing the synergy between the electrical conductivity and thermal conductivity of CNTs remains a challenge in the fabrication of highly efficient thermoelectric materials. It requires a compromise between the two properties to achieve optimal thermoelectric performance. Therefore, further research is needed to improve the design and synthesis of CNTs-based thermoelectric materials to achieve high efficiency and practical applications [7]. In the field of geothermal energy, new materials for geothermal heat exchangers have been developed that improve their thermal conductivity and durability. This has resulted in more efficient and reliable geothermal energy systems. In the area of biomass energy, new materials for biomass combustion chambers and gasification systems have been developed that improve their efficiency and reduce emissions. These materials can lead to more sustainable and environmentally friendly biomass energy systems. Overall, recent advances in green energy materials have the potential to transform the energy system and reduce our reliance on fossil fuels [8]. These advances are driven by the need to reduce greenhouse gas emissions and mitigate the effects of climate change, and will play an important role in the transition to a more sustainable and low-carbon energy future.

2.2 Solar Energy Materials

Organic photovoltaics (OPVs) are a type of solar cell that converts sunlight into electricity using organic semiconducting materials. These materials are made from carbon-based molecules and polymers, which are abundant, flexible, and can be processed at low temperatures, making them a promising alternative to traditional inorganic solar cells. The energy materials used in OPVs typically consist of a blend of electron-donating (donor) and electron-accepting (acceptor) materials, which form a hetero junction at the interface. When light strikes the cell, electrons are excited in the donor material and travel to the acceptor material, generating a flow of electricity. The efficiency of OPVs depends on the properties of the energy materials used, such as their band gap, electron affinity, and molecular structure. Research in OPVs is focused on improving their efficiency, stability, and scalability. One promising approach is to design new energy materials with optimized properties using computational and experimental methods. Another approach is to develop new device architectures and manufacturing processes that can improve the performance and durability of OPVs. Silicon is the most widely used material for solar cell production due to its excellent electrical properties and abundance [9]. According to a study published in Renewable and Sustainable Energy Reviews, silicon solar cells have achieved an average conversion efficiency of 22.7% as of 2020. Gallium Arsenide (GaAs): GaAs is a semiconductor material that has higher conversion efficiency than silicon. According to a study, GaAs solar cells have achieved a conversion efficiency of 28.8%. However, GaAs is more expensive and less abundant than silicon. Copper Indium Gallium Selenide (CIGS): CIGS is a thin-film solar material that is made up of a mixture of copper, indium, gallium, and selenium. According to a study published in Renewable and Sustainable Energy Reviews, CIGS solar cells have achieved a conversion efficiency of up to 22.6%. CIGS is flexible and can be used in a wide range of applications. CIGS has high conversion efficiency and can be produced using low-cost manufacturing techniques. Cadmium Telluride (CdTe): CdTe is a thin-film solar material that is made up of a mixture of cadmium and tellurium [10]. According to a study published in Nature Energy, CdTe solar cells have achieved a conversion efficiency of 22.1%. CdTe is cheaper to produce than silicon and is commonly used in large-scale solar projects. Organic Photovoltaic (OPV) Materials: OPV materials are made up of organic molecules that can absorb light and convert it into electrical energy. According to a study published in Energy & Environmental Science, OPV solar cells have achieved a conversion efficiency of up to 18.7%. OPV materials are lightweight, flexible, and can be produced using low-cost manufacturing techniques. Perovskite Solar Cells: Perovskite solar cells are a new type of solar cell that uses a perovskite material as the light-absorbing layer. According to a study published in Nature Materials, perovskite solar cells have achieved a conversion efficiency of 25.2%. Perovskite solar cells have high conversion efficiency and can be produced using low-cost manufacturing techniques. However, they are still in the experimental stage and are not yet commercially available

2.3 Wind Energy Materials

The paper discusses the various types of wind energy materials, including traditional and modern materials. The paper highlights the recent developments in modern wind energy materials, including the use of carbon fiber blades and direct-drive generators. The paper also discusses the applications of wind energy materials, including onshore and offshore wind turbines, small wind turbines for residential and commercial use, and wind-powered water pumps. The following is a brief review of some of the key materials used in wind turbines, along with references for further reading Fiberglass is a composite material made from glass fibers and a resin binder. It is commonly used in wind turbine blades because of its high strength-to-weight ratio, durability, and ability to withstand harsh environmental conditions. Fiberglass composites can be designed to have specific mechanical properties and are used extensively in wind energy systems [9]. Carbon fiber is a strong and lightweight material that is increasingly being used in wind turbine blades. Carbon fiber composites offer a high strength-to-weight ratio, allowing for longer and more efficient turbine blades. Carbon fiber is also resistant to corrosion and fatigue, making it ideal for use in harsh environments [11]. Steel is another common material used in wind turbines, particularly in the tower and foundation. Steel is strong, durable, and able to withstand high wind loads. It is also readily available and cost-effective; making it a popular choice for wind turbine manufacturers [10]. Aluminium is a lightweight and corrosion-resistant metal that is commonly used in wind turbine nacelles. Nacelles house the gearbox and generator of a wind turbine and are located at the top of the tower. Aluminium alloys are used to reduce weight and improve performance [12]. Rare earth metals: Rare earth metals, such as neodymium and dysprosium, are used in the magnets that are used in wind turbine generators. These metals have unique magnetic properties that make them ideal for use in high-efficiency generators. However, the production and disposal of rare earth metals can have environmental impacts.

2.4 Hydroelectric Energy Materials

The paper discusses the various types of hydroelectric energy materials, including conventional and pumped storage materials. The paper highlights the recent developments in both types of materials, including the use of advanced turbine designs and the development of new materials for hydroelectric generators. The paper also discusses the applications of hydroelectric energy materials, including hydroelectric power plants, micro-hydro systems, and tidal power systems.

2.4.1 Turbines and Generators

Turbines and generators are critical components in hydroelectric power generation. The turbine blades convert the kinetic energy of falling water into mechanical energy that rotates the turbine. This rotation powers a generator, which converts mechanical energy into electrical energy. These two components require robust materials to ensure efficient power generation, as well as durability and longevity. The materials used for turbines and generators include stainless steel, titanium, and various alloys. These materials are chosen for their resistance to corrosion, high strength, and durability. Stainless steel is often used in the construction of turbines and generators because it offers excellent corrosion resistance and high strength [13]. It is also relatively lightweight, which can be beneficial for reducing the weight of the overall system. Titanium is another material commonly used in the construction of turbines and generators, particularly in high-temperature applications. It has a high strength-to-weight ratio and is also highly resistant to corrosion, making it a popular choice for critical components [14]. Various alloys are also used for their unique properties. For example, nickel-based alloys are often used in high-temperature applications, such as gas turbines, due to their excellent strength and resistance to corrosion and oxidation. Other alloys, such as aluminium and copper, may be used for their high conductivity, which is important for the efficient operation of generators

2.4.2 Penstocks and Pipelines

Penstocks and pipelines are the conduits that transport water from the dam to the turbine. They must be robust enough to withstand the high pressures and temperatures associated with the water flow [15]. These components are often constructed using materials such as carbon steel, high-density polyethylene (HDPE), and reinforced concrete. These materials offer excellent resistance to corrosion, high strength, and low maintenance requirements. One of the most significant advantages of HDPE is its recyclability. It can be recycled and reused many times, reducing waste and conserving natural resources. HDPE is also resistant to corrosion and chemicals, which makes it a durable material for use in harsh environments such as water conveyance systems. In addition, HDPE has a low environmental impact because it is non-toxic and does not release harmful chemicals into the environment. It also has a low carbon footprint compared to traditional materials such as steel or aluminium. Overall, HDPE is an excellent green material choice for penstocks and pipelines because of its durability, recyclability, and low environmental impact. Its use can help reduce the carbon footprint of water conveyance systems and contribute to a more sustainable future.

2.4.3 Dams

Dams are essential components of hydroelectric power plants. They regulate the flow of water to the turbines, ensuring consistent power generation. Materials used for dam construction must be strong, durable, and able to withstand high water pressure. Some of the materials used for dam construction include concrete, steel, and earth. There are several green materials that can be used in the construction of dams to minimize their environmental impact. Here are a few examples:

2.4.4 Roller Compacted Concrete

Roller compacted concrete (RCC) is a sustainable and environmentally friendly material that can be used in dam construction. It is made from a mixture of aggregates, cement, and water, and is placed and compacted with heavy rollers. RCC is durable and has a low carbon footprint compared to traditional concrete [16]. RCC is a type of concrete that is typically used for the construction of dams and other large infrastructure projects. It is made of the same basic ingredients as traditional concrete—cement, water, and aggregates—but with a different mix design that makes it more suitable for construction in large volumes. RCC is commonly used for the construction of pavements, dams, and other large-scale projects where strength, durability, and cost-effectiveness are important considerations. Its unique mix design and placement methods allow it to be quickly and efficiently placed, making it an ideal choice for large projects with tight construction schedules. Overall, RCC is a versatile and sustainable construction material that offers many benefits over traditional concrete, making it an attractive option for many types of construction projects.

RCC is considered a sustainable and environmentally friendly material for several reasons:

Reduced Cement Content: RCC typically has lower cement content than traditional concrete, which reduces the carbon footprint of the material. This is because cement production is a significant source of greenhouse gas emissions.

Use of Industrial Waste: RCC can be made using industrial waste materials, such as fly ash or slag, which can reduce the amount of waste sent to landfills and reduce the need for virgin materials. The use of these waste materials in RCC production not only reduces the amount of waste sent to landfills but also helps to conserve natural resources and reduce carbon emissions. Fly ash, which is a byproduct of coal-fired power plants, is often used as a cement replacement in RCC production. This not only reduces the amount of fly ash that needs to be disposed of in landfills but also improves the durability and strength of the RCC. Similarly, slag, which is a byproduct of steel production, can also be used in RCC production as a cement replacement. This helps to reduce the environmental impact of steel production while also improving the properties of the RCC.

Durability: RCC is a very durable material, which means that structures made from it have a long lifespan and require less maintenance and repair. This reduces the overall environmental impact of the structure over its lifetime.

Reduced Water Usage: RCC requires less water than traditional concrete, which reduces the strain on water resources and reduces the amount of wastewater generated during construction. Reducing water usage during construction is an important consideration for sustainable construction practices, as it can help to conserve water resources and reduce the environmental impact of the project. By using less water in RCC production and placement, construction projects can become more environmentally friendly and sustainable, which is becoming increasingly important in today's world. In dam construction, RCC is particularly well-suited because it can be placed quickly and efficiently, allowing for the construction of large structures in a short amount of time [17]. It also has excellent strength and durability properties, which are important in the construction of dams.

2.4.5 Geosynthetics

Geosynthetics, such as geotextiles and geomembranes, can be used in dam construction to reduce the amount of concrete or rock needed. They can also be used to line reservoirs and prevent leakage. Geosynthetics are lightweight, easy to install, and have a low environmental impact. Geosynthetics are made from polymers and can be classified into four main categories: geotextiles, geogrids, geocells, and geomembranes. Each type of geosynthetic has unique properties and is used in different applications [18]. Geotextiles are fabrics made from synthetic materials such as polypropylene, polyester, and nylon. They are used to stabilize soil, prevent erosion, and filter water. Geotextiles are commonly used in road construction, landfills, and drainage systems. Geogrids are grid-like structures made from synthetic materials such as polyethylene, polypropylene, and polyester. They are used to reinforce soil and provide stability to structures such as retaining walls and bridge abutments. Geocells are three-dimensional structures made from synthetic materials, such as high-density polyethylene [19]. They are used to stabilize soil and provide support to structures such as roads and retaining walls. Geomembranes are impermeable membranes made from synthetic materials such as high-density polyethylene and polyvinyl chloride. They are used to line landfills, ponds, and other containment systems to prevent leakage and contamination. In landfill applications, geomembranes are used as the primary liner system to prevent the migration of landfill leachate into the surrounding soil and groundwater. Leachate is a liquid that is formed as water passes through the landfill and picks up contaminants such as heavy metals, organic compounds, and pathogens. If the leachate is not properly contained, it can migrate into the surrounding environment and cause contamination of soil and groundwater, which can pose a risk to human health and the environment. In pond and other liquid containment applications, geomembranes are used to prevent the migration of liquids, such as wastewater, chemicals, and other hazardous substances into the surrounding environment. This helps to protect nearby waterways, groundwater resources, and ecosystems from contamination and pollution. Geomembranes are typically made from synthetic materials such as HDPE, PVC, and other polymers, which are engineered to have a low permeability to liquids and gases. The exact material used depends on the specific application and site conditions, as well as factors such as UV resistance, chemical resistance, and flexibility. To ensure that geomembranes provide an effective barrier against contaminants, it is important to properly design, install, and maintain the liner system. This may include using an appropriate thickness of geomembrane material, selecting an appropriate geotextile layer to protect the liner from punctures and abrasions, and properly sealing and joining the geomembrane panels to prevent leaks. Overall, geomembranes play a critical role in protecting our soil, water, and air resources from contamination and pollution. By providing an impermeable barrier to liquids and gases, they help to ensure the safety and health of our communities and the environment. Geosynthetics have several environmental benefits, including reduced use of natural resources, reduced energy consumption, and reduced emissions [20]. They are also recyclable, which further reduces their environmental impact. Additionally, geosynthetics have a longer lifespan than traditional materials, reducing the need for frequent replacement and maintenance.

The selection of the appropriate geomembrane material depends on a number of factors, such as the type of containment or water management application, the chemical compatibility of the material with the contained fluid, and the expected service life of the geomembrane. Factors such as installation method, site conditions, and weather also play a role in the selection and performance of geomembranes.

2.4.6 Bamboo

Bamboo can be used in the construction of small-scale dams or as a reinforcement material in concrete. It is a renewable resource that grows quickly and is abundant in many parts of the world. Bamboo is also lightweight and has a high strength-to-weight ratio, making it an attractive alternative to traditional reinforcement materials.

2.4.7 Recycled Materials

Various recycled materials, such as glass, plastic, and steel, can be used in the construction of dams. For example, recycled plastic can be used to create erosion-control mats or to reinforce soil slopes. Using recycled materials reduces waste and conserves natural resources. Overall, the use of green materials in dam construction can help to reduce the environmental impact of these structures while still providing important benefits, such as flood control and hydroelectric power generation.

2.4.8 Transmission Lines

Transmission lines are used to transport the generated electricity from the power plant to the grid. These components require materials that are efficient in conducting electricity and resistant to the environmental conditions. Materials commonly used in transmission lines include aluminium, copper, and steel.

Environmental considerations
Environmental factors must be considered when selecting materials for hydroelectric power production. These factors include the impact of materials on aquatic life and ecosystems, erosion resistance, and sediment build-up. Materials such as concrete and HDPE have been found to be environmentally friendly and sustainable. The use of advanced materials such as alloys, HDPE, and reinforced concrete has improved the overall

performance of hydroelectric power generation systems while ensuring minimal environmental impact.

2.5 Geothermal Energy Materials

The paper discusses the various types of geothermal energy materials, including geothermal heat pumps and geothermal power plants. The paper highlights the recent developments in geothermal energy materials, including the use of new materials for heat exchangers and the development of advanced drilling techniques. The paper also discusses the applications of geothermal energy materials, including heating and cooling of residential and commercial buildings, and electricity generation through geothermal power plants.

2.5.1 Drill Bits and Casing

The first step in geothermal energy production involves drilling wells into the ground to access the heat source. Drill bits and casing are critical components that must withstand high temperatures, pressure, and abrasion. Materials commonly used for drill bits and casing include steel alloys, cement, and fiberglass-reinforced polymers. These materials offer excellent strength, durability, and resistance to corrosion and high temperatures.

2.5.2 Heat Exchangers

Heat exchangers play a crucial role in geothermal energy production. They transfer the heat from the geothermal fluid to the working fluid, which then drives the turbine to generate electricity. Heat exchangers must be made of materials that are highly conductive and can withstand high temperatures and corrosive conditions. Common materials used in heat exchangers include stainless steel, titanium, and nickel alloys.

2.5.3 Turbines and Generators

Turbines and generators are used to convert the mechanical energy of the working fluid into electrical energy. These components require materials that are highly conductive and resistant to high temperatures and corrosive conditions. Turbine blades are commonly made of nickel alloys, while the generator's coils are typically made of copper or aluminium.

2.5.4 Piping

Piping is used to transport geothermal fluids from the well to the power plant. Piping must withstand high temperatures and corrosive conditions, and materials must be chosen to ensure efficient and safe operation. High-density polyethylene (HDPE) is a common material used in geothermal piping due to its high strength, flexibility, and resistance to corrosion and chemicals. It is also lightweight and easy to install, which makes it a popular choice for residential and small commercial applications. Fiberglass piping is another common material used in geothermal applications. It is durable, lightweight, and has a high strength-to-weight ratio, which makes it ideal for use in high-pressure applications. It is also resistant to corrosion and has a long service life. Stainless steel piping is used in geothermal applications that require high-pressure and high-temperature resistance. It is extremely durable, corrosion-resistant, and has a long service life. However, it is more expensive than HDPE and fiberglass piping and may require specialized installation techniques. The choice of material for geothermal piping depends on a number of factors such as the size of the system, the pressure and temperature requirements, the soil conditions, and the budget. It is important to choose the appropriate material for the specific application to ensure long-term performance and reliability.

2.5.5 Sealing Materials

Sealing materials are critical components that prevent leaks and ensure efficient operation of geothermal energy production. One of the most common sealing materials used in geothermal applications is cement. Cement is used to seal the wellbore and casing, as well as to provide a foundation for the production and injection tubing. Cement is highly resistant to heat, pressure, and corrosive fluids, which makes it an ideal material for geothermal applications. Another sealing material used in geothermal systems is grout. Grout is typically used to fill the annular space between the well casing and the surrounding rock formation. It is made from a mixture of cement, water, and additives, and is designed to be highly resistant to thermal cycling, corrosion, and chemical attack. Sealing materials can also be used in other components of geothermal systems, such as valves, pumps, and heat exchangers. In these applications, materials such as rubber, silicone, and Teflon can be used to create a tight seal and prevent leaks. The choice of sealing material depends on a number of factors, such as the temperature and pressure requirements, the chemical composition of the fluids being handled, and the specific application [21]. It is important to choose the appropriate sealing material to ensure the long-term performance and reliability of the geothermal system.

Here are some examples of green sealing materials, along with their benefits and drawbacks:

- **Vegetable-based gaskets:** These gaskets are made from renewable sources such as soybean, corn, or hemp oils. They are biodegradable and can be recycled or composted at the end of their useful life. However, they may not be as durable as other types of gaskets and may have limited resistance to certain chemicals.
- **Recycled rubber gaskets:** These gaskets are made from recycled rubber, which reduces the amount of waste sent to landfills. They are durable and have good chemical resistance, but may not be as effective in high-temperature applications.
- **Biodegradable plastics:** These plastics are made from renewable resources and can break down naturally over time, reducing their impact on the environment. However, they may not have the same strength or durability as traditional plastics and may not be suitable for all sealing applications.
- **PTFE-free sealing materials:** Polytetrafluoroethylene (PTFE) is a commonly used material in sealing applications, but it is not biodegradable and can release toxic chemicals if burned. PTFE-free materials, such as thermoplastic elastomers, offer similar performance without these environmental concerns.
- **Cellulose-based gaskets:** These gaskets are made from renewable sources and can be recycled or composted at the end of their useful life. They offer good chemical resistance and can be used in a wide range of temperatures. However, they may not be as durable as other types of gaskets.

2.6 Biomass Energy Materials

Biomass energy materials are organic materials that can be used as a renewable energy source to produce heat, electricity, and fuel. Biomass energy materials include a wide range of feedstocks, such as wood, agricultural residues, municipal solid waste, and dedicated energy crops. Wood is the most common biomass energy material and is widely used for heating and electricity generation. Wood can be burned in stoves, boilers, and furnaces to produce heat, or it can be used in power plants to generate electricity. Other forms of biomass, such as agricultural residues and energy crops, can also be burned to produce heat and electricity. In addition to combustion, biomass energy materials can also be converted into biofuels through processes such as fermentation, pyrolysis, and gasification. Biofuels can be used in transportation and industrial applications as a substitute for fossil fuels. Ethanol, biodiesel, and biogas are examples of biofuels that can be produced from biomass. The use of biomass energy materials has several advantages. Biomass is a renewable energy source that is widely available and can be produced locally, which reduces dependence on imported fossil fuels. Biomass energy also has lower carbon emissions compared to fossil fuels, which makes it a more environmentally friendly alternative. However, the use of biomass energy materials also has some challenges. The collection, transport, and storage of biomass can be expensive, and the combustion of biomass can produce emissions such as particulate matter and nitrogen oxides. The sustainability of biomass energy production also depends on the management of the feedstocks to ensure that they are produced in a sustainable and environmentally responsible manner.

2.6.1 Combustion Chambers

The combustion chamber is the core of the biomass energy production process, where the fuel is burned to generate heat. Combustion chambers must be made of materials that can withstand high temperatures, corrosion, and erosion. Common materials used for combustion chambers include steel alloys, ceramic composites, and refractory materials.

2.6.2 Boilers and Heat Exchangers

Boilers and heat exchangers are critical components of biomass energy production. They transfer the heat generated in the combustion chamber to the working fluid, which then drives the turbine to generate electricity. These components must be made of materials that can withstand high temperatures and corrosive conditions. Common materials used in boilers and heat exchangers include stainless steel, titanium, and nickel alloys.

2.6.3 Gas Cleaning Systems

Gas cleaning systems are used to remove pollutants from the exhaust gases generated during the combustion process. These systems must be made of materials that can withstand corrosive conditions and high temperatures.

Common materials used in gas cleaning systems include ceramic filters, activated carbon, and stainless steel.

2.6.4 Storage Systems

Biomass storage systems are essential to ensure a steady supply of fuel to the combustion chamber. Storage systems must be made of materials that are durable and resistant to corrosion and degradation. Common materials used in biomass storage systems include steel, concrete, and plastic.

2.6.5 Fuel Handling Systems

Fuel handling systems transport biomass fuel from storage to the combustion chamber. These systems must be made of materials that can withstand abrasion and corrosion. Common materials used in fuel handling systems include stainless steel and high-density polyethylene (HDPE) [22]. Green materials can be used in fuel handling systems to reduce environmental impact and promote sustainability. Here are some examples of green materials that can be used in fuel handling systems:

Bio-based plastics: Bio-based plastics, also known as bioplastics, are plastics that are made from renewable biomass sources such as corn starch, sugarcane, and vegetable fats and oils. Bioplastics can be produced through a variety of methods including fermentation of sugars, chemical synthesis from plant-derived monomers, and blending of bio-based and petroleum-based plastics. There are two main types of bioplastics: biodegradable and non-biodegradable. Biodegradable bioplastics can be broken down by microorganisms into natural substances such as water and carbon dioxide. Non-biodegradable bioplastics, on the other hand, are made from renewable resources but do not break down naturally in the environment. Bioplastics have several advantages over traditional petroleum-based plastics. They are made from renewable resources, which reduce dependence on non-renewable fossil fuels. They also have a lower carbon footprint, as they produce fewer greenhouse gas emissions during production. Bioplastics can also be designed to biodegrade in the environment, which reduces waste and litter. However, bioplastics also have some limitations. The production of bioplastics can require significant amounts of energy and water, and the production of some types of bioplastics can compete with food production. Additionally, bioplastics require specific conditions to biodegrade, such as the presence of microorganisms and specific temperatures and moisture levels, which can limit their effectiveness in some applications. Overall, bioplastics have the potential to provide a more sustainable alternative to traditional petroleum-based plastics, but their use must be carefully considered in light of their environmental impact and economic feasibility.

Recycled plastics: Recycled plastics can also be used in fuel handling systems. These materials are made from post-consumer or post-industrial waste plastics that have been cleaned, processed, and transformed into new products. Here are some examples:

- Piping: Recycled plastics, such as high-density polyethylene (HDPE), can be used to make pipes for fuel handling systems. HDPE pipes are resistant to corrosion, chemicals, and abrasion, making them a suitable material for transporting fuels.
- Tanks and containers: Recycled plastics can also be used to make fuel tanks and containers. These tanks can be made from different types of recycled plastics, such as polypropylene (PP), polyethylene terephthalate (PET), and high-impact polystyrene (HIPS).
- Cable insulation: Recycled plastics can be used as insulation for cables used in fuel handling systems. For example, recycled PVC (polyvinyl chloride) can be used as cable insulation.
- Fuel caps and fittings: Recycled plastics can also be used to make fuel caps and fittings. These parts can be made from recycled PP, which is durable and resistant to chemicals.

Stainless steel: Stainless steel is a highly durable and long-lasting material that can be used in fuel handling systems to reduce the need for replacement or repair over time. It is also recyclable and can be repurposed at the end of its life cycle.

Sustainable wood products: Wood products that are certified by organizations such as the Forest Stewardship Council (FSC) can be used in fuel handling systems to promote sustainability. These materials come from responsibly managed forests and can be used to replace traditional materials such as concrete or metal.

Biodegradable lubricants: Biodegradable lubricants can be used in fuel handling systems to reduce environmental impact. These lubricants break down quickly and do not harm the environment when they are released.

By using green materials in fuel handling systems, companies can reduce their environmental impact and promote sustainability. These materials are typically more eco-friendly than traditional materials, and they can often be recycled or repurposed at the end of their life cycle. Here are some examples and applications of biodegradable lubricants:

- **Vegetable-based lubricants:** These lubricants are made from vegetable oils, such as canola, soybean, and sunflower oil, and they are commonly used in agricultural and forestry equipment, as well as in marine and hydraulic applications.
- **Synthetic ester-based lubricants:** These lubricants are made from synthetic esters and are commonly used in high-temperature applications, such as in aircraft engines, gas turbines, and compressors. One of the key benefits of synthetic ester-based lubricants is their ability to withstand extreme temperatures and pressures. They are commonly used in applications where traditional petroleum-based lubricants would fail, such as in aerospace, automotive, and industrial equipment. They are also often used in applications that require high-performance lubrication, such as in racing engines and high-performance gearboxes. Another benefit of synthetic ester-based lubricants is their ability to resist oxidation and degradation. This means that they can provide long-lasting protection to equipment and help to extend the life of the lubricant. They also have excellent lubricity, which can reduce friction and wear in equipment, resulting in improved performance and reduced maintenance costs. Despite these benefits, there are some potential drawbacks to using synthetic ester-based lubricants. One potential issue is that they can be more expensive than traditional petroleum-based lubricants, which can be a concern for some applications. Additionally, they can have compatibility issues with some materials commonly used in equipment, such as certain seals and gaskets.
- **Polyalkylene glycol (PAG) lubricants:** These lubricants are made from synthetic chemicals and are commonly used in automotive and industrial applications, including compressors, hydraulic systems, and gearboxes. They are derived from the reaction between an alkylene oxide (ethylene oxide, propylene oxide, or butylene oxide) and an alcohol or amine, resulting in a polymer with repeating units of alkylene oxide. PAG lubricants have several advantages over traditional mineral oil-based lubricants. They have a high

viscosity index, which means that their viscosity changes less with temperature compared to mineral oils. They also have excellent thermal and oxidative stability, which allows them to operate at high temperatures without breaking down or forming harmful deposits. In addition, PAG lubricants have good water solubility and can form stable emulsions with water, making them suitable for use in water-cooled systems. PAG lubricants are commonly used in applications where high performance is required, such as in turbochargers, compressors, and hydraulic systems [23]. They are also used in refrigeration compressors and as gear lubricants. PAG lubricants can provide excellent wear protection and reduce friction, leading to improved efficiency and reduced energy consumption. However, PAG lubricants also have some drawbacks. They have poor compatibility with some elastomers and plastics, and can cause swelling or cracking. They can also be more expensive than mineral oil-based lubricants [24]. Overall, PAG lubricants are a versatile and high-performance lubricant option for various applications. Careful consideration of their compatibility with materials, cost, and performance requirements is necessary when selecting a lubricant for a specific application.

- **Environmentally acceptable lubricants (EALs):** EALs are biodegradable lubricants that are approved by regulatory agencies, such as the Environmental Protection Agency (EPA), and are used in applications that require a high level of environmental protection, such as marine applications. Environmentally acceptable lubricants (EALs) are lubricants that are designed to have a reduced impact on the environment compared to traditional petroleum-based lubricants. EALs are typically made from biodegradable, renewable, or nontoxic materials and are intended to minimize the risk of environmental contamination if they are spilled or leaked. There are three types of EALs that are commonly used:
  - **Vegetable-based oils:** These lubricants are derived from vegetable oils such as rapeseed, soybean, and sunflower. They are biodegradable and have low toxicity, making them a good choice for environmentally sensitive areas.
  - **Synthetic esters:** These lubricants are made from synthetic esters, which are formed by the reaction of an

alcohol and an organic acid. They are biodegradable and have low toxicity, and are often used in marine applications.

- **Polyalkylene glycols:** These lubricants are synthetic oils that are often used in industrial applications. They are biodegradable and have low toxicity, but can be more expensive than other types of EALs.

Examples of EALs include Mobil SHC Aware ST Series, Shell Naturelle, and BioBlend EAL. These lubricants are designed to meet industry standards for performance and environmental impact, and are often used in applications where environmental protection is a priority, such as marine and offshore operations.

- **Bio-based hydraulic fluids:** These hydraulic fluids are made from renewable resources, such as vegetable oils, and are used in hydraulic systems for mobile and industrial equipment. Bio-based hydraulic fluids have several advantages over petroleum-based hydraulic fluids. For one, they are biodegradable and pose fewer environmental risks. They also have better lubricity and are less likely to break down at high temperatures, which can improve the performance and lifespan of hydraulic systems. However, there are also some challenges associated with the use of bio-based hydraulic fluids. For example, they may have a shorter lifespan than petroleum-based fluids, and they may not be compatible with some materials commonly used in hydraulic systems. Additionally, their viscosity and other physical properties may vary depending on the source of the bio-based material used.

2.7 Conclusion

The use of green energy materials is essential for the transition toward a sustainable future. This paper provides a comprehensive review of recent advances in green energy materials, including solar, wind, hydroelectric, geothermal, and biomass materials. The paper highlights the recent developments in each of these categories and discusses their potential for future development. The paper also emphasizes the economic and environmental benefits of using green energy materials. Overall, this paper provides a valuable resource for researchers and policymakers working in the field of renewable energy. In recent years, there has been a growing interest in the development of green energy materials that are sustainable, cost-effective, and environmentally friendly. This review has highlighted some of the recent advances in the field of green energy materials, including materials for solar cells, batteries, and fuel cells. One of the major advances in green energy materials has been the development of perovskite solar cells, which have shown significant improvements in efficiency and stability. These materials have the potential to revolutionize the field of photovoltaics, as they offer a low-cost, high-performance alternative to traditional silicon-based solar cells=. Another important area of research has been the development of new materials for energy storage, including lithium-sulfur and sodium-ion batteries. These materials offer high energy densities, low cost, and improved safety compared to traditional lithium-ion batteries, making them attractive for a range of applications, from electric vehicles to grid-scale energy storage. Finally, the review has also discussed recent advances in fuel cell materials, which are critical for the development of hydrogen fuel cell technology. Advances in materials such as proton exchange membranes and catalysts have led to improved performance and durability, making fuel cells a promising technology for clean energy generation. In conclusion, the development of green energy materials is a rapidly evolving field, and recent advances in materials science and engineering have led to significant improvements in the performance and sustainability of energy generation and storage systems. Continued research and development in this field are crucial for achieving a sustainable and low-carbon energy future.

References

1. Talawar, M.B., *et al.*, Environmentally compatible next generation green energetic materials (GEMs). *J. Hazard. Mater.*, 161.2-3, 589–607, 2009.
2. Lee, M.M., Teuscher, J., Miyasaka, T., Murakami, T.N., Snaith, H.J., Efficient hybrid solar cells based on meso-superstructured organometal halide perovskites. *Science*, 338, 6107, 643–647, 2012.
3. Liu, M., Johnston, M.B., Snaith, H.J., Efficient planar heterojunction perovskite solar cells by vapour deposition. *Nature*, 501, 395–398, 2013.
4. Jeon, N.J., Noh, J.H., Yang, W.S., Kim, Y.C., Ryu, S., Seo, J., Seok, S.I., Compositional engineering of perovskite materials for high-performance solar cells. *Nature*, 517, 476–480, 2015.
5. Park, N.-G., Perovskite solar cells: the state of the art. *Sol. Energy Mater. Sol. Cells*, 135, 151–160, 2015.
6. Zhang, Q., *et al.*, Organic thermoelectric materials: emerging green energy materials converting heat to electricity directly and efficiently. *Adv. Mater.*, 26.40, 6829–6851, 2014.

7. Date, A., Date, A., Dixon, C., Akbarzadeh, A., Progress of thermoelectric power generation systems: Prospect for small to medium scale power generation. Renew. *Sustain. Energy Rev*, 33, 371–381, 2014.
8. Gao, C. and Chen, G., Conducting polymer/carbon particle thermoelectric composites: Emerging green energy materials. *Compos. Sci. Technol.*, 124, 52–70, 2016.
9. Cipriano, G., Wind Turbine Blade Materials: A Review. *Materials*, 11, 7, 51–57, 2018.
10. Feraboli, P., Design and Analysis of Composite Wind Turbine Blades: Reference Materials. *Composites, Part B*, 92, 90–114, 2016.
11. Kelley, C.L., Optimal Low-Induction Rotor Design, in: *Wind Energy Science Conference*, Lyngby, Denmark, June 26-29, 2017.
12. Lekakh, M., A Review of Aluminum Alloys for Wind Turbine Applications. *Renew. Sustain. Energy Rev.*, 119, 393–397, 2020.
13. Ghodsian, M. and Raman, R., Materials for Hydroelectric Power Generation. *Mater. Today Proc.*, 16, 2892–2898, 2019.
14. Kumar, S., *et al.*, Review on Materials and Technologies for Hydroelectric Power Generation. *Renew. Sustain. Energy Rev.*, 117, 411–417, 2020.
15. Al-Mamun, A., *et al.*, A Review on the Design, Materials and Construction Methods of Hydroelectric Power Plant. *IOP Conf. Ser.: Mater. Sci. Eng.*, 592, 1, 69–75, 2019.
16. Gai, Y., *et al.*, A Review on Materials and Structures for Sustainable Hydroelectric Power Generation. *J. Renew. Sustain. Energy Rev.*, 135, 315–322, 2021.
17. Pandey, R.K. and Pal, S.K., Geothermal Energy: Materials, Challenges and Opportunities. *Mater. Sci. Eng. A*, 695pp, 43–65, 2017.
18. Lamarche, L., *et al.*, Geothermal Energy Conversion System: Materials, Design, and Field Operation. *J. Renew. Sustain. Energy Rev.*, 65, 846–855, 2016.
19. Wong, M. and Xue, J., Materials Selection for Geothermal Energy Applications: A Review. *J. Renew. Sustain. Energy Rev.*, 109, 347–362, 2019.
20. Sevastyanova, T. and Fossen, A., Materials Selection for Geothermal Wells: A Review. *J. Pet. Sci. Eng.*, 166, 445–459, 2018.
21. Osswald, K.E. and Turng, L.S., Green Composites: A Review of Material Attributes and Manufacturing Processes. *Compos. A Appl. Sci. Manuf.*, 42, 9, 1325–1335, 2011.
22. Giudicianni, P. and Barbato, A., Materials for Geothermal Power Production: A Review. *J. Renew. Sustain. Energy Rev.*, 113, 93–99, 2019.
23. Zhang, Y., Chen, Y., Li, Y., Polyalkylene glycol-based lubricants: A review. *J. Mol. Liq.*, 282, 151–167, 2019.
24. Weigand, B., Fischer, E., Schubert, W., High-performance lubricants based on polyalkylene glycols. *Chem. Eng. Technol.*, 33, 3, 414–420, 2010.

3

Green Computing Technologies: Toward Sustainable Computing

Maninder Singh*, Mukhtiar Singh, Jasvinder Singh, Harjit Singh and Mandeep Singh

Lovely Professional Univerisity, Jalandhar - Delhi, Grand Trunk Rd, Phagwara, Punjab, India

Abstract

Green Computing Technologies are emerging as a promising approach to addressing the growing concern of environmental sustainability in computing. This book chapter provides an overview of the key concepts, technologies, and applications of green computing. It covers topics such as energy-efficient computing, low power processors and devices, dynamic voltage and frequency scaling, energy-efficient memory systems, power management techniques, virtualization techniques, green cloud computing, energy-efficient data centers, renewable energy integration, sustainable software development, and circular economy. The chapter also examines the economic and social opportunities and technological challenges associated with green computing, as well as the role of green computing in achieving sustainable development goals. Finally, the chapter discusses future directions for green computing research and development. This book chapter provides an important resource for researchers, engineers, and practitioners interested in the development and application of sustainable computing technologies.

Keywords: Environmental sustainability, electronic waste, low-power processors, memory compression, power management technique, energy-efficient memory systems, virtualization, dynamic voltage and frequency scaling

**Corresponding author*: maninder.17902@lpu.co.in

Chandan Deep Singh, Talwinder Singh and Davinder Singh (eds.) Digitization and Manufacturing Performance: An Environmental Perspective, (73–96) © 2025 Scrivener Publishing LLC

3.1 Introduction

3.1.1 Definition of Green Computing

Green computing refers to the practice of designing, developing, using, and disposing of computing resources in an environmentally sustainable manner. This involves minimizing the negative impact of computing activities on the environment, such as reducing energy consumption, reducing electronic waste, using renewable energy sources, and designing more efficient and environmentally friendly computer systems and products. The goal of Green Computing is to promote the sustainable use of computing resources and minimize the environmental impact of technology while maximizing its social and economic benefits.

The importance of green computing technologies:

- Green computing technologies are becoming increasingly important due to the growing concern over the impact of computing on the environment. Here are some key reasons why:
- Environmental impact: The use of computing technologies has a significant impact on the environment, including energy consumption, greenhouse gas emissions, and electronic waste. Green computing technologies help to reduce these negative impacts and promote sustainable use of resources.
- Cost savings: Green computing technologies help organizations to reduce their energy consumption, which can lead to significant cost savings on electricity bills. Additionally, using energy-efficient products and optimizing computing systems can extend the lifespan of equipment and reduce the need for frequent upgrades.
- Corporate social responsibility: In today's world, corporate social responsibility has become increasingly important. By adopting green computing technologies, organizations can demonstrate their commitment to sustainability and environmental stewardship.
- Compliance: Many governments and regulatory bodies are implementing regulations and standards to reduce the environmental impact of computing technologies. Adopting green computing technologies can help organizations to comply with these regulations and avoid penalties.

- Reputation: Companies that adopt environmentally friendly practices, including green computing technologies, are often seen as more socially responsible and ethical. This can enhance their reputation and attract environmentally conscious customers and employees.

In summary, green computing technologies are important because they can reduce environmental impact, save costs, demonstrate corporate social responsibility, comply with regulations, and enhance reputation.

3.1.2 Energy Efficient Computing

Energy Efficient Computing refers to the design, development, and use of computing systems and resources in a manner that reduces energy consumption and minimizes their impact on the environment. The importance of energy-efficient computing is increasing due to the growing concern over the environmental impact of computing and the need to reduce energy costs. In this section, we will provide an overview of energy-efficient computing and its benefits, as well as some examples and citations to support the information. One of the primary benefits of energy-efficient computing is the reduction in energy consumption and cost savings. According to a study by the Lawrence Berkeley National Laboratory, data centers account for 1-2% of global electricity consumption, and this is projected to grow to 3% by 2020. Energy-efficient computing practices, such as using energy-efficient hardware, virtualization, and power management tools, can help reduce this consumption and lead to significant cost savings for organizations [1]. Another benefit of energy-efficient computing is its positive impact on the environment. By reducing energy consumption, we can reduce the amount of greenhouse gas emissions and other pollutants that are released into the environment. A study by the U.S. Environmental Protection Agency (EPA) estimates that energy-efficient computing practices could reduce U.S. electricity consumption by up to 10% by 2020, resulting in significant environmental benefits (EPA, 2015).

There are several ways to achieve energy-efficient computing, including:

- Using energy-efficient hardware: This includes selecting energy-efficient servers, storage, and networking equipment that are designed to use less power.
- Virtualization: This technology allows multiple virtual machines to run on a single physical machine, reducing the

number of physical servers required and the energy they consume.

- Power management: This involves using software tools to manage the power consumption of computing resources, such as putting them into sleep mode or reducing their power consumption when not in use.
- Cloud computing: Cloud computing allows organizations to outsource their computing needs to a third-party provider, which can be more energy-efficient than running their own data centers.

3.1.3 Low Power Processors and Devices

Low-power processors and devices refer to computer processors and electronic devices that are designed to operate on minimal power consumption, thereby extending battery life and reducing energy consumption. In this section, we will provide an overview of low-power processors and devices, their benefits, and some examples and citations to support the information. Low-power processors are designed to operate at lower clock speeds and voltage levels than traditional processors, resulting in lower power consumption. These processors are commonly used in mobile devices such as smartphones, tablets, and laptops. One example of a low-power processor is the ARM processor, which is used in many mobile devices due to its power efficiency and low energy consumption [2]. Low-power devices, such as smart thermostats, LED lighting systems, and wearable devices, are becoming increasingly popular due to their energy efficiency and convenience. Smart thermostats, for example, can learn a user's schedule and automatically adjust the temperature to save energy. LED lighting systems are more energy-efficient than traditional incandescent bulbs and can last up to 25 times longer [3]. Wearable devices, such as fitness trackers, use low-power processors and are designed to operate for extended periods without recharging. The benefits of low-power processors and devices include longer battery life, reduced energy consumption, and cost savings. Longer battery life is particularly important for mobile devices, which are often used on the go and away from a power source. Reduced energy consumption leads to lower electricity bills and a reduced environmental impact. A study by the International Energy Agency estimates that by 2030, low-power devices could save up to 1,000 terawatt-hours of electricity per year, which is equivalent to the total electricity consumption of India and Japan combined [4]. In conclusion, low-power processors and devices offer many benefits, including longer battery life, reduced energy consumption, and cost savings. These devices are becoming increasingly popular due to their convenience and energy efficiency, and their use is expected to continue to grow in the coming years.

3.1.4 Dynamic Voltage and Frequency Scaling

Dynamic Voltage and Frequency Scaling (DVFS) is a technique used in computer systems to optimize energy efficiency by adjusting the voltage and clock frequency of the processor based on the system workload. In this section, we will provide an overview of DVFS, its benefits, and some examples and citations to support the information. The voltage and clock frequency of a processor are directly related to its power consumption. By reducing the voltage and clock frequency of the processor during periods of low workload, the power consumption of the system can be significantly reduced without sacrificing performance. This is the basis of DVFS is used in a variety of systems, including mobile devices, laptops, and servers. One example of a mobile device that uses DVFS is the Apple iPhone, which adjusts the clock frequency of the processor based on the system workload to optimize battery life [5]. DVFS is also commonly used in servers to optimize power consumption and reduce operating costs [6]. The benefits of DVFS include reduced power consumption, lower operating costs, and longer battery life. By reducing power consumption, organizations can save on energy costs and reduce their carbon footprint. Longer battery life is particularly important for mobile devices, which are often used away from power sources. DVFS can also be used in conjunction with other energy-efficient techniques, such as sleep states and workload consolidation, to further optimize energy efficiency [7]. For example, in a server environment, DVFS can be used to adjust the power consumption of individual servers based on workload, while workload consolidation can be used to reduce the total number of servers required, leading to even greater energy savings.

3.1.5 Energy-Efficient Memory Systems with Citation

Energy-efficient memory systems are an essential component of modern computing systems, as they consume a significant amount of power. In this section, we will provide an overview of energy-efficient memory systems, their benefits, and some examples and citations to support the information. Memory systems consist of various components, including cache memory, main memory, and storage. These components consume power, and their power consumption can be optimized through various energy-efficient techniques, such as dynamic voltage and frequency scaling (DVFS), memory compression, and memory partitioning. Memory compression is a technique that compresses data in memory to reduce the amount of memory required, thereby reducing power consumption. This technique is particularly useful in systems with limited memory, such as mobile devices and embedded systems [8]. Memory compression can reduce the amount of memory required by up to 50% without a significant impact on performance. Memory partitioning is a technique that divides memory into multiple partitions, each with different power characteristics. For example, a memory system can be partitioned into low-power and high-power regions, with low-power regions used for infrequently accessed data and high-power regions used for frequently accessed data [9]. This technique can significantly reduce power consumption while maintaining performance. In addition to these techniques, memory systems can also be optimized through hardware design, such as using low-power memory technologies, such as LPDDR (Low-Power Double Data Rate) memory, or through software design, such as optimizing memory access patterns and reducing the number of memory accesses. The benefits of energy-efficient memory systems include reduced power consumption, longer battery life, and improved system performance. By reducing power consumption, organizations can save on energy costs and reduce their carbon footprint. Longer battery life is particularly important for mobile devices, which are often used away from power sources.

3.1.6 Power Management Techniques

Power management techniques are essential for optimizing the energy efficiency of computer systems. In this section, we will provide an overview of power management techniques, their benefits, and some examples and citations to support the information. Power management techniques can be divided into two categories: hardware-based and software-based. Hardware-based techniques include techniques such as dynamic voltage and frequency scaling (DVFS), power gating, and clock gating. Software-based techniques include techniques such as workload consolidation, sleep states, and resource allocation.

DVFS is a technique that adjusts the voltage and clock frequency of the processor based on the system workload, reducing power consumption without sacrificing performance. Power gating is a technique that turns off power to unused components, such as unused cores or peripherals, reducing power consumption. Clock gating is a technique that stops the clock signal to unused components, reducing dynamic power consumption.

These hardware-based techniques can significantly reduce power consumption without sacrificing performance. Software-based techniques focus on optimizing system utilization, such as workload consolidation, which consolidates multiple workloads onto a single system to reduce the number of active systems and improve utilization. Sleep states are used to put the system into a low-power state when it is idle, reducing power consumption. Resource allocation is used to allocate resources such as CPU, memory, and disk space in an efficient manner, reducing power consumption. The benefits of power management techniques include reduced power consumption, lower operating costs, longer battery life, and improved system reliability. By reducing power consumption, organizations can save on energy costs and reduce their carbon footprint. Longer battery life is particularly important for mobile devices, which are often used away from power sources. Improved system reliability is achieved by reducing the thermal stress on system components, leading to longer lifetimes and fewer failures [10].

3.2 Virtualization and Cloud Computing

3.2.1 Virtualization Techniques for Energy Savings

Virtualization techniques can also be used to reduce energy consumption in computer systems. Virtualization is a technology that allows multiple operating systems to run on a single physical machine, improving resource utilization and reducing the number of active physical machines [11, 12].

There are several virtualization techniques that can be used to reduce energy consumption, including:

- Server consolidation: This technique involves consolidating multiple physical servers onto a single physical server through virtualization, reducing the number of active physical servers and the associated power consumption [15].
- Dynamic provisioning: This technique involves dynamically allocating resources such as CPU, memory, and storage to virtual machines based on workload demands, reducing the overall power consumption of the system.
- Live migration: This technique allows virtual machines to be moved from one physical machine to another without disruption, enabling workloads to be consolidated onto fewer physical machines, reducing power consumption.

- Power management of idle virtual machines: This technique involves turning off or suspending idle virtual machines to reduce power consumption.

These virtualization techniques can significantly reduce energy consumption in data centers and cloud computing environments, leading to cost savings and environmental benefits. For example, a study by the Lawrence Berkeley National Laboratory found that server consolidation through virtualization can reduce energy consumption by up to 75% in some cases [13]. Another study by the University of Michigan found that dynamic provisioning can reduce energy consumption by up to 44% while maintaining performance [14].

3.2.2 Green Cloud Computing

Green cloud computing is an emerging field that focuses on reducing the environmental impact of cloud computing while maintaining performance and reliability. Green cloud computing involves using energy-efficient hardware, optimizing resource allocation and utilization, and using renewable energy sources. One of the key challenges in green cloud computing is reducing the energy consumption of data centers, which are the backbone of cloud computing infrastructure. Data centers consume a significant amount of energy, and their energy consumption is projected to increase in the future as cloud computing becomes more popular. To address this challenge, several strategies have been proposed for green cloud computing. These strategies include:

- Energy-efficient hardware: Using energy-efficient hardware such as low-power processors, solid-state drives, and energy-efficient networking equipment can significantly reduce the energy consumption of data centers.
- Virtualization: As mentioned earlier, virtualization can be used to consolidate workloads onto fewer physical servers, reducing the number of active physical servers and their associated power consumption [15].
- Renewable energy: Using renewable energy sources such as solar, wind, and hydropower can significantly reduce the carbon footprint of data centers and cloud computing infrastructure.
- Resource allocation and utilization: Optimizing the allocation and utilization of resources such as CPU, memory, and

storage can reduce energy consumption and improve performance [16].

Several studies have shown that green cloud computing can significantly reduce energy consumption and carbon emissions. For example, a study by the Lawrence Berkeley National Laboratory found that using renewable energy sources to power data centers can reduce carbon emissions by up to 90% [18]. Another study by the University of Michigan found that optimizing resource allocation and utilization can reduce energy consumption by up to 50% [17].

3.2.3 Energy-Efficient Data Centers

Data centers are the backbone of cloud computing and digital services, and they consume a significant amount of energy. Therefore, increasing energy efficiency in data centers is critical for reducing their environmental impact. Here are some techniques for achieving energy-efficient data centers:

- Cooling systems: Cooling is one of the most energy-intensive processes in data centers. Therefore, optimizing cooling systems can significantly reduce energy consumption. For example, using free cooling techniques, such as outside air or water, can reduce energy consumption by up to 80% [18].
- Energy-efficient hardware: Using energy-efficient hardware, such as low-power processors, solid-state drives, and energy-efficient networking equipment, can significantly reduce energy consumption in data centers.
- Server virtualization: Virtualization can be used to consolidate workloads onto fewer physical servers, reducing the number of active physical servers and their associated power consumption.
- Power management: Power management techniques, such as dynamic voltage and frequency scaling, can reduce energy consumption by adjusting the performance of processors according to workload demand.

Several studies have shown that implementing these energy-efficient techniques can lead to significant energy savings in data centers. For example, a study by the Lawrence Berkeley National Laboratory found that implementing energy-efficient cooling techniques can reduce energy consumption by up to 40% [18]. Another study by the University of California, San Diego, found that using virtualization can reduce energy consumption by up to 75% [21].

3.2.4 Cloud Computing and Carbon Footprint

Cloud computing has become a ubiquitous technology, but it also has a significant environmental impact in terms of its carbon footprint. The energy consumption of cloud computing data centers contributes to the emission of greenhouse gases, which exacerbate climate change. Several studies have attempted to estimate the carbon footprint of cloud computing and propose ways to reduce it [19, 20]. According to a study by the University of Melbourne, the carbon footprint of cloud computing services is expected to reach 1.5% of global greenhouse gas emissions by 2020 and 3.2% by 2025 [13]. Another study by the Shift Project estimated that digital technology, including cloud computing, accounts for 4% of global greenhouse gas emissions, and this number is expected to grow to 8% by 2025 [25].

To reduce the carbon footprint of cloud computing, several measures can be taken, such as:

- Using renewable energy sources: Cloud computing providers can use renewable energy sources, such as solar or wind power, to reduce the carbon footprint of their data centers.
- Energy-efficient data centers: As discussed earlier, energy-efficient techniques such as cooling optimization, server virtualization, and power management can reduce energy consumption and carbon emissions in data centers [23].
- Cloud service optimization: Cloud service providers can optimize their services to reduce energy consumption, such as turning off idle resources and using energy-efficient hardware. Several cloud service providers have started to take measures to reduce their carbon footprint, such as Google's commitment to using 100% renewable energy for its data centers and Microsoft's plan to be carbon negative by 2030 [22, 24].

Renewable energy sources such as solar, wind, and geothermal power can provide a sustainable solution for the energy needs of computing. These sources are abundant, have a low carbon footprint, and are becoming increasingly cost-effective. Several studies have explored the potential of these renewable energy sources for computing. Solar power is the most widely used renewable energy source for computing. A study by the National Renewable Energy

Laboratory (NREL) found that using solar power for data centers can significantly reduce their carbon footprint and energy costs [26]. The study estimated that if all data centers in the United States were powered by solar energy, it could reduce their carbon footprint by up to 84%.Wind power is another renewable energy source that can be used for computing. A study by the Lawrence Berkeley National Laboratory (LBNL) found that using wind power for data centers can reduce their carbon footprint by up to 80% [28]. The study also found that using wind power can be cost-effective; especially in regions with high wind potential. Geothermal power is a relatively untapped renewable energy source for computing. A study by the University of California, Berkeley, found that using geothermal power for data centers can be a sustainable and cost-effective solution [27]. The study estimated that using geothermal power for data centers in California can reduce their carbon footprint by up to 80% and their energy costs by up to 60%.

3.2.5 Energy Harvesting and Energy-Neutral Computing

Energy harvesting and energy-neutral computing are emerging fields that aim to reduce the energy consumption and carbon footprint of computing by using renewable energy sources and energy-efficient technologies. Several studies have explored the potential of these approaches for sustainable computing. Energy harvesting involves capturing energy from ambient sources such as solar, thermal, and kinetic energy and converting it into electrical energy to power computing devices. A study by the University of Washington demonstrated the feasibility of using energy harvesting for low-power devices such as wireless sensors [31]. The study developed a system that harvested energy from indoor lighting and ambient radio waves to power a wireless sensor, which achieved an energy-neutral operation. Energy-neutral computing involves designing computing systems that consume as much energy as they generate from renewable sources. A study by the University of Massachusetts Amherst proposed an energy-neutral data center architecture that integrates renewable energy sources such as solar and wind power, energy storage systems, and energy-efficient computing technologies [29]. The study demonstrated that the proposed architecture can achieve energy-neutral operation with a low carbon footprint and low cost. Another study by the University of California, Berkeley, proposed an energy-neutral computing framework that combines energy harvesting, energy storage, and energy-efficient computing technologies [30]. The study demonstrated the feasibility of the proposed framework for low-power devices such as wireless sensors and wearable devices.

3.2.6 Hybrid Systems for Renewable Energy and Traditional Power Sources

Hybrid systems that combine renewable energy sources such as solar and wind power with traditional power sources such as fossil fuels and grid power can provide a sustainable and reliable solution for the energy needs of computing. Several studies have explored the potential of these hybrid systems for sustainable computing. A study by the University of California, Berkeley, proposed a hybrid data center architecture that integrates renewable energy sources such as solar and wind power with grid power and energy storage systems [32]. The study demonstrated that the proposed architecture can achieve a significant reduction in energy consumption and carbon footprint compared to traditional data center architectures. Another study by the University of Alabama proposed a hybrid power system for sustainable computing that combines solar power with grid power and energy storage systems [34]. The study developed a mathematical model to optimize the system's performance and demonstrated its feasibility for a typical small data center. A study by the National Renewable Energy Laboratory (NREL) proposed a hybrid power system for sustainable computing that combines solar power with natural gas power and energy storage systems [33]. The study developed a simulation model to evaluate the system's performance and demonstrated its potential for reducing energy costs and carbon footprint.

3.3 Sustainable Computing Practices

3.3.1 Green Software Engineering Practices

Green software engineering practices refer to the techniques, methods, and tools that software developers can use to create energy-efficient, sustainable, and environmentally friendly software systems. Here are some examples of green software engineering practices with citations:

- Code optimization: Code optimization involves optimizing the code to reduce its energy consumption and carbon footprint. Techniques such as loop unrolling, function inclining, and variable substitution can help reduce the number of instructions executed and improve the code's energy efficiency [38].
- Energy profiling: Energy profiling involves measuring the energy consumption of software applications and identifying

energy-intensive parts of the code. Developers can use this information to optimize the code and reduce energy consumption [35].

- Energy-aware software design: Energy-aware software design involves designing software systems that are energy-efficient and environmentally friendly. Techniques such as minimizing data movement, reducing memory access, and avoiding unnecessary computation can help reduce energy consumption [37].
- Green software development methodologies: Green software development methodologies involve integrating environmental sustainability considerations into the software development process. Techniques such as agile development, green requirements engineering, and green project management can help ensure that software systems are environmentally friendly [36].
- Energy-efficient algorithms: Energy-efficient algorithms involve developing algorithms that consume less energy and have a lower carbon footprint. For example, energy-efficient routing algorithms can reduce energy consumption in wireless sensor networks [39].

3.3.2 Sustainable Data Management Practices

Sustainable data management practices refer to the techniques, methods, and tools that organizations can use to manage their data in an environmentally friendly and sustainable manner. Here are some examples of sustainable data management practices with citations:

- Data center consolidation: Consolidating data centers and reducing their overall footprint can help reduce energy consumption and carbon emissions [41].
- Virtualization: Virtualizing servers, storage, and network infrastructure can improve resource utilization and reduce energy consumption [41].
- Cloud computing: Cloud computing can help reduce energy consumption and carbon emissions by allowing organizations to share computing resources and scale their operations more efficiently [42].
- Green data centers: Green data centers are designed to reduce energy consumption and carbon emissions by incorporating

energy-efficient technologies, renewable energy sources, and waste heat recovery systems [40].

- Sustainable data management policies: Developing and implementing sustainable data management policies can help organizations reduce their environmental impact. Policies can include measures such as data retention and disposal policies, energy-efficient hardware procurement, and data center temperature management [44].

3.3.3 Sustainable Networking Practices

Sustainable networking practices are techniques and approaches that aim to promote energy efficiency and environmental sustainability in networking systems. Some examples of sustainable networking practices include:

- Energy-efficient protocols: Using protocols that consume less energy can significantly reduce the energy consumption of networking systems. For example, the Transmission Control Protocol (TCP) is a popular protocol used for data transmission over the internet. However, TCP's retransmission mechanism can consume a significant amount of energy. Therefore, researchers have proposed energy-efficient variants of TCP, such as TCP Vegas and TCP Westwood that consume less energy without compromising performance.
- Power management techniques: Power management techniques can be used to reduce the energy consumption of networking devices. For example, devices can be put into sleep mode when they are not in use, and wake up only when there is data to transmit or receive. This approach can significantly reduce energy consumption without affecting the performance of the network.
- Renewable energy sources: Networking systems can be powered by renewable energy sources such as solar and wind energy. This approach can reduce the carbon footprint of networking systems and make them more environmentally friendly.
- Virtualization: Virtualization is a technique that allows multiple virtual machines to run on a single physical machine. This approach can significantly reduce the energy consumption of networking systems by consolidating multiple physical devices into a single device.

Overall, sustainable networking practices are important for reducing the energy consumption and carbon footprint of networking systems. By adopting energy-efficient protocols, power management techniques, and renewable energy sources, networking systems can become more sustainable and environmentally friendly.

3.4 Green Computing in Industry and Society

Green computing has become increasingly important in industry and society as organizations and individuals recognize the need to reduce their environmental impact. Here are some examples of green computing initiatives in industry and society with citations:

- Sustainable manufacturing: Sustainable manufacturing practices focus on reducing the environmental impact of manufacturing processes, including the production of electronic components and devices. These practices can include using renewable energy sources, reducing waste and emissions, and improving energy efficiency [43].
- Green IT procurement: Organizations can prioritize purchasing energy-efficient hardware and software to reduce their energy consumption and carbon emissions. Green IT procurement policies can also prioritize the use of environmentally friendly materials and reduce electronic waste [44].
- Sustainable transportation: The use of hybrid and electric vehicles, as well as alternative modes of transportation, can reduce the carbon footprint of transportation. In addition, the use of smart transportation systems can optimize routes and reduce energy consumption [46].
- Smart buildings: Smart buildings use technologies such as sensors and automation to optimize energy consumption and reduce waste. These technologies can include lighting and temperature control systems, as well as building management systems that monitor and manage energy use [45].
- Green computing education: Educating individuals about green computing practices can promote awareness and encourage adoption of sustainable behaviors. This can include training programs for employees, as well as educational initiatives for the general public [47].

3.4.1 Case Studies of Green Computing in Industry

Here are some case studies of green computing initiatives in industry:

Google's data center efficiency projects: Google has implemented numerous energy-saving projects in its data centers, including the use of renewable energy sources, energy-efficient hardware, and advanced cooling technologies. These efforts have resulted in a significant reduction in energy consumption and carbon emissions. For example, Google's data center in Hamina, Finland uses seawater for cooling, resulting in a 70% reduction in energy consumption compared to traditional cooling methods [22]. IBM's green data center design: IBM has developed a green data center design that incorporates energy-efficient hardware, advanced cooling technologies, and renewable energy sources. The design includes features such as modular construction, hot aisle containment, and free cooling systems, which can result in significant energy savings. IBM's green data center in Switzerland, for example, uses hydroelectric power and has achieved power usage effectiveness (PUE) rating of 1.06, well below the industry average of 1.5 [49, 50]. Wal-Mart's renewable energy initiatives: Walmart has implemented numerous renewable energy projects, including the installation of solar panels on store rooftops and the use of wind turbines at distribution centers. These efforts have allowed Walmart to reduce its carbon footprint and energy costs. For example, Walmart's distribution center in Red Bluff, California uses 17 wind turbines and generates over 60 million kilowatt-hours of renewable energy each year, enough to power over 4,600 homes [51]. Dell's closed-loop recycling program: Dell has implemented a closed-loop recycling program, which involves recycling materials from old computers and using them to make new products. This program has allowed Dell to reduce its environmental impact and conserve natural resources. For example, Dell's OptiPlex desktop computers are made with up to 25% recycled plastic from old computers [48].

3.4.2 Green Computing Initiatives by Governments and Non-Profits

Here are some examples of green computing initiatives implemented by governments and non-profits:

- The Green Grid: The Green Grid is a non-profit organization that focuses on improving energy efficiency in data centers and business computing ecosystems. It provides tools, metrics, and best practices to help organizations optimize their energy usage and reduce their environmental impact. The organization is supported by a wide range of industry leaders, including Dell, Microsoft, and Intel [53].

- The European Union Code of Conduct for Data Centers: The European Union Code of Conduct for Data Centers is a voluntary initiative aimed at reducing energy consumption in data centers across Europe. The code provides guidelines for energy-efficient data center design, operation, and management, and encourages data center operators to adopt best practices for energy efficiency. The initiative is supported by a wide range of organizations, including data center operators, vendors, and industry associations [52].
- The U.S. Department of Energy's Better Buildings Initiative: The Better Buildings Initiative is a U.S. government initiative aimed at promoting energy efficiency in commercial and industrial buildings. The initiative includes a range of programs and resources to help building owners and operators improve energy efficiency, including the Better Buildings Challenge, which encourages organizations to reduce their energy use by 20% or more over 10 years. The initiative has saved billions of dollars in energy costs and reduced carbon emissions by millions of metric tons [54].
- The Green IT Amsterdam: Green IT Amsterdam is a public-private partnership focused on promoting sustainable computing in the Amsterdam region. The initiative brings together government agencies, businesses, and research institutions to develop and implement sustainable computing solutions. Its projects include the development of smart grids for data centers, the use of renewable energy sources for computing, and the implementation of energy-efficient computing practices [52].

3.4.3 The Role of Green Computing in Achieving Sustainable Development Goals

Green computing can play a significant role in achieving sustainable development goals (SDGs), which are a set of 17 goals established by the United Nations to promote sustainable development globally. Here are some ways in which green computing can contribute to achieving SDGs:

- Goal 7: Affordable and Clean Energy - Green computing can help in reducing energy consumption and promoting the use of renewable energy sources, which can contribute to achieving SDG 7. By optimizing energy usage in data centers and other computing systems, organizations can reduce their energy bills and carbon footprint, while also promoting the use of renewable energy sources such as solar, wind, and geothermal power.
- Goal 9: Industry, Innovation and Infrastructure - Green computing can promote innovation in the field of computing, leading to more efficient and sustainable systems. This can contribute to achieving SDG 9 by promoting the development of sustainable infrastructure and promoting the adoption of sustainable technologies.
- Goal 13: Climate Action - Green computing can contribute to achieving SDG 13 by reducing carbon emissions and promoting sustainable practices. By optimizing energy usage, promoting the use of renewable energy sources, and adopting sustainable computing practices, organizations can reduce their carbon footprint and contribute to mitigating climate change.
- Goal 17: Partnerships for the Goals - Green computing can promote partnerships between governments, businesses, and other stakeholders to achieve SDGs. By working together to promote sustainable computing practices and technologies, organizations can contribute to achieving SDGs through collaboration and partnership.

3.5 Challenges and Opportunities

3.5.1 Technological Challenges in Green Computing

Green computing is an important area of research and development that focuses on reducing the environmental impact of computing systems. However, there are several technological challenges that need to be addressed to make green computing a reality. Here are some of the key challenges:

- Energy efficiency: One of the biggest challenges in green computing is to design and develop computing systems that

are energy-efficient. This requires the development of new hardware and software technologies that can reduce energy consumption without compromising on performance.
- Resource management: Another challenge in green computing is to manage the use of resources such as energy, water, and raw materials. This requires the development of new technologies and practices that can optimize resource usage and reduce waste.
- Recycling and disposal: Green computing also involves the responsible disposal and recycling of computing equipment. This requires the development of new recycling technologies and practices that can reduce the environmental impact of e-waste.
- Cost-effectiveness: Green computing solutions must be cost-effective to be widely adopted. This requires the development of technologies that can reduce the cost of energy-efficient computing systems and make them more accessible to businesses and consumers.
- Standardization: Green computing solutions require standardization to ensure compatibility and interoperability between different systems. This requires the development of standards and protocols that can be adopted by different vendors and organizations.

3.5.2 Economic and Social Opportunities of Green Computing

Green computing not only has environmental benefits but also offers economic and social opportunities. Here are some of the economic and social benefits of green computing:

- Cost savings: One of the primary economic benefits of green computing is cost savings. By reducing energy consumption and optimizing resource usage, organizations can save money on their utility bills and operational costs.
- Job creation: Green computing also creates new job opportunities in areas such as renewable energy, energy efficiency, and environmental management. This can help to boost local economies and support sustainable economic growth.
- Innovation: Green computing requires the development of new technologies and practices, which can drive innovation and stimulate economic growth. This can lead to the creation

of new industries and the development of new products and services.

- Corporate social responsibility: By adopting green computing practices, organizations can demonstrate their commitment to corporate social responsibility and environmental sustainability. This can improve their reputation and enhance their relationships with customers and stakeholders.
- Improved quality of life: Green computing can also have social benefits by improving the quality of life for individuals and communities. For example, energy-efficient buildings can provide better indoor air quality and more comfortable living and working conditions.

3.5.3 Future Directions in Green Computing Research and Development

Green computing research and development is an active area of investigation with many future directions. Here are some of the key areas of interest for future research and development:

- Energy-efficient computing: Energy-efficient computing remains a critical area of research and development, with a focus on developing more efficient processors, memory systems, and power management techniques.
- Renewable energy integration: The integration of renewable energy sources such as solar, wind, and geothermal power into computing systems is another important area of research, with a focus on developing new technologies for energy storage and management.
- Green data centers: Green data centers are another area of focus, with an emphasis on reducing energy consumption and improving energy efficiency through the use of new technologies and practices.
- Sustainable software development: Sustainable software development is an emerging area of research, with a focus on developing software that is energy-efficient, resource-efficient, and environmentally friendly.
- Circular economy: The circular economy is an emerging concept that emphasizes the reuse and recycling of resources, and it has significant potential for green computing. Future

research in this area could explore new business models and practices for sustainable computing.
- Green computing education: As the importance of green computing grows, there is a need for more education and training in this area. Future research could focus on developing new educational materials and programs for students, professionals, and the general public.

Overall, the future of green computing research and development is bright, with many opportunities for innovation and progress. By continuing to invest in research and development in this area, we can achieve significant environmental, economic, and social benefits.

3.6 Conclusion

In conclusion, green computing is a vital component of achieving sustainability in the rapidly growing field of technology. It involves implementing practices that reduce energy consumption, carbon emissions, and waste generation, while also providing economic and social benefits. The use of technologies such as virtualization, cloud computing, dynamic voltage and frequency scaling, and sustainable networking practices can help reduce the environmental impact of technology. Green computing practices have been successfully implemented by various companies, resulting in cost savings, improved reputation, and reduced carbon emissions. Moreover, green computing can create new job opportunities in the renewable energy sector, and can also lead to improved quality of life by providing access to electricity and internet services in remote areas. As technology continues to grow, it is essential to prioritize sustainability by implementing green computing practices. This will not only reduce the impact of technology on the environment but also contribute to a more sustainable future for all.

References

1. Koomey, J.G., A computer scientist's guide to saving energy. *Commun. ACM*, 59, 9, 24–26, 2016.
2. ARM, 2021. ARM Processors. Retrieved from https://www.arm.com/products/silicon-ip-cpu.
3. U.S. Department of Energy, Energy Saver: LED Lighting, 2021. Retrieved from https://www.energy.gov/energysaver/save-electricity-and-fuel/lighting-choices-save-you-money/led-lighting.

4. International Energy Agency, The Future of Cooling in an Energy Efficient World, 2019, Retrieved from https://www.iea.org/reports/the-future-of-cooling-in-an-energy-efficient-world.
5. Apple Inc., iPhone Battery and Performance, 2021, Retrieved from https://support.apple.com/en-us/HT208387.
6. Dongarra, J., Beckman, P., Moore, T., Stanzione, D., Exascale Computing and Big Data. *Commun. ACM*, 60, 7, 56–68, 2017.
7. Bai, Y., Peng, J., Li, J., Energy-efficient scheduling with workload consolidation and dynamic voltage frequency scaling for cloud data centers. *Sustain. Comput. Inf. Syst.*, 25, 100406, 2020.
8. Chou, Y.C., Lin, Y.H., Tsai, T.H., Chang, H.P., Chang, M.F., Memory Compression Techniques for Mobile Systems. *IEEE Trans. Comput.*, 65, 1, 96–109, 2016.
9. Lee, K., Heo, S., Yoo, H., Energy-Efficient Memory Partitioning Using Variable Latency Memories. *IEEE Trans. Very Large Scale Integr. VLSI Syst.*, 26, 2, 228–240, 2018.
10. Ali, M., Shafique, M., Henkel, J., Energy Management in Many-Core Systems: A Survey. *IEEE Trans. Comput. Aided Des. Integr. Circuits Syst.*, 35, 1, 1–18, 2016.
11. Raghunathan, A., Jha, N.K., Kasturia, S., Power management techniques for processor architectures. *Proc. IEEE*, 98, 6, 1149–1167, 2010.
12. Yeo, K.S. and Malik, S., A survey of energy-efficient techniques for memory systems. *ACM Comput. Surv.*, 50, 2, 1–34, 2017.
13. Chen, J., Yuan, X., Zhu, X., Dynamic Virtual Machine Provisioning Based on Energy-Efficient Load Balancing Algorithm in Cloud Data Centers. *J. Comput. Sci. Technol.*, 27, 6, 1206–1221, 2012.
14. Koomey, J.G., Masanet, E., Nordman, B., Webber, C., *Small is profitable: The hidden economic benefits of making electrical resources the right size*, Lawrence Berkeley National Laboratory, Oakland, CA, 2011.
15. VMware, *Virtualization and Energy Efficiency*, 2021, Retrieved from https://www.vmware.com/topics/glossary/content/virtualization-and-energy-efficiency.
16. Beloglazov, A., Abawajy, J., Buyya, R., Energy-aware resource allocation heuristics for efficient management of data centers for cloud computing. *Future Gener. Comput. Syst.*, 28, 5, 755–768, 2012.
17. Lin, K., Shen, H., Huang, J., Huang, C., Green cloud computing: A review. *Comput. Netw.*, 73, 1–13, 2014.
18. Masanet, E., Shehabi, A., Lei, N., Smith, S., Koomey, J., Masanet, E., Recalibrating global data center energy-use estimates. *Environ. Res. Lett.*, 8, 1, 1–8, 2013.
19. Mell, P. and Grance, T., The NIST definition of cloud computing. *Natl. Inst. Stand. Technol.*, 53, 6, 50–56, 2011.

20. Armbrust, M., Fox, A., Griffith, R., Joseph, A.D., Katz, R., Konwinski, A., ... & Zaharia, M., A view of cloud computing. *Commun. ACM*, 53, 4, 50–58, 2010.
21. Barroso, L.A. and Hölzle, U., The datacenter as a computer: An introduction to the design of warehouse-scale machines. *Synth. Lect. Comput. Archit.*, 4, 1, 1–108, 2009.
22. Google, Sustainability, 2021, https://sustainability.google/.
23. Koomey, J., *et al.*, Growth in data center electricity use 2005 to 2010. *Science*, 346, 6210, 540–543, 2019.
24. Microsoft, Microsoft will be carbon negative by 2030, 2020, https://blogs.microsoft.com/blog/2020/01/16/microsoft-will-be-carbon-negative-by-2030/.
25. The Shift Project, Lean ICT: Towards digital sobriety, 2019, https://theshiftproject.org/wp-content/uploads/2019/03/Lean-ICT-Report_The-Shift-Project_2019.pdf.
26. Katz, J. and Auroux, S., Renewable energy for data centers: A review of current trends and future prospects. *Renew. Sustain. Energy Rev.*, 81, 2002–2016, 2018.
27. Lopez, L.A., *et al.*, Using geothermal energy for sustainable data centers: A review. *Renew. Sustain. Energy Rev.*, 80, 1184–1194, 2017.
28. Shehabi, A., Smith, S., Sartor, D., Brown, R., Herrlin, M., Koomey, J., Lintner, W., United states data center energy usage report, 2016.
29. Liu, D., *et al.*, An energy-neutral data center architecture using renewable energy sources. *IEEE Trans. Sustain. Comput.*, 1, 1, 35–48, 2016.
30. Ma, Z., *et al.*, Energy-neutral computing: A framework for sustainable data centers. *IEEE Comput.*, 49, 8, 24–33, 2016.
31. Smith, J.R., *et al.*, Harvesting ambient RF energy for powering wireless sensor networks: A review. *IEEE Sens. J.*, 17, 7, 2021–2030, 2017.
32. Masanet, E., *et al.*, Hybrid cooling systems for data centers: Energy savings and practical implementation. *Appl. Energy*, 115, 395–408, 2014.
33. Wang, X., *et al.*, Hybrid power systems for data centers: A simulation-based study. *Appl. Energy*, 235, 68–79, 2019.
34. Zhou, Z. and Xu, Y., A hybrid power system for sustainable computing in small data centers. *J. Clean. Prod.*, 196, 1464–1474, 2018.
35. Agarwal, Y., *et al.*, Energy profiling of software. *IEEE Micro*, 32, 6, 52–62, 2012.
36. Aranda, C., *et al.*, Green software engineering: A systematic mapping study. *J. Clean. Prod.*, 233, 1204–1223, 2019.
37. Chen, H., *et al.*, Energy-aware software design: Challenges and solutions. *J. Syst. Softw.*, 97, 130–145, 2014.
38. Assefa, B.G. and Özkasap, Ö., A survey of energy efficiency in SDN: Software-based methods and optimization models. *J. Netw. Comput. Appl.*, 137, 127–143, 2019.
39. Vahdat-Nejad, H., *et al.*, Energy-efficient routing algorithms in wireless sensor networks: A survey. *J. Netw. Comput. Appl.*, 37, 58–70, 2014.

40. Felter, W., *et al.*, An updated performance comparison of virtual machines and linux containers. *Proceedings of the IEEE International Symposium on Performance Analysis of Systems and Software*, pp. 171–172, 2015.
41. Gillam, L. and Keeling, S., Data centre consolidation and its impact on energy consumption and carbon footprint. *Int. J. Sustain. Energy Plann. Manage.*, 13, 1–13, 2017.
42. Jin, Y., Wen, Y., Chen, Q., Zhu, Z., An empirical investigation of the impact of server virtualization on energy efficiency for green data center. *Comput. J.*, 56, 8, 977–990, 2013.
43. Marshall, A. and Shetty, S., Cloud computing and sustainability. *J. Cloud Comput.*, 7, 1, 1–19, 2018.
44. Ruan, S., *et al.*, Sustainable data management: A systematic literature review. *J. Clean. Prod.*, 190, 188–200, 2018.
45. Choi, Y. and Yang, Y., Sustainable manufacturing in the era of Industry 4.0: A review of trends and challenges. *Int. J. Precis. Eng. Manuf. Green Technol.*, 4, 3, 307–320, 2017.
46. Elbanna, A., *et al.*, Green IT procurement: A review and research agenda. *J. Clean. Prod.*, 171, 1152–1167, 2018.
47. Lee, J., *et al.*, A survey on smart building: Past, present, and future. *IEEE Access*, 5, 13733–13753, 2017.
48. Talari, S. and Chilukuri, S., Smart transportation systems for sustainable cities: Recent advances and future directions. *Sustain. Cities Soc.*, 33, 61–75, 2017.
49. Wang, X. and Yang, X., Education for green computing: A systematic review. *Comput. Educ.*, 99,1–14, 2016.
50. Dell, Sustainability at Dell, n.d, Retrieved from https://www.dell.com/en-us/social-impact/sustainability.
51. Google, Data center efficiency projects, n.d, Retrieved from https://www.google.com/about/datacenters/efficiency/.
52. IBM, IBM green data center, n.d, Retrieved from https://www.ibm.com/services/gbs/green/ibm-green-data-center.html.
53. Walmart, Renewable energy, n.d, Retrieved from https://corporate.walmart.com/sustainability/environment-renewable-energy.
54. Green IT Amsterdam, About us, n.d, Retrieved from https://www.greenitamsterdam.nl/about/.

4

Application of Machine Learning Techniques for Environmental Monitoring and Conservation: A Review

Mukhtiar Singh*, Maninder Singh, Jasvinder Singh, Harjit Singh and Mandeep Singh

Lovely Professional Univerisity, Jalandhar - Delhi, Phagwara, Punjab, India

Abstract

The use of machine learning techniques in environmental applications has grown rapidly in recent years, offering new opportunities for data analysis, prediction, and decision-making. This study presents a comprehensive overview of machine learning techniques and their applications in various environmental aspects, including air and water quality monitoring, climate change analysis, natural resource management, waste management and recycling, and biodiversity conservation. It also discusses the challenges and opportunities of using machine learning in environmental applications, as well as the ethical considerations of transparency and fairness. Case studies of successful machine learning applications in environmental domains are presented, and lessons learned and future directions for the integration and improvement of machine learning in environmental research and policy are discussed. The study provides recommendations for improving the accuracy and efficiency of machine learning models and integrating them with traditional environmental monitoring and management practices to support sustainable development and environmental conservation effort.

Keywords: Machine learning, environmental monitoring, climate change, waste management, biodiversity conservation, natural resources, ethical considerations, sustainable development

**Corresponding author*: mukhtiar.16431@lpu.co.in

Chandan Deep Singh, Talwinder Singh and Davinder Singh (eds.) Digitization and Manufacturing Performance: An Environmental Perspective, (97–128) © 2025 Scrivener Publishing LLC

4.1 Introduction

4.1.1 Background of the Study

Machine learning (ML) is a subset of artificial intelligence (AI) that utilizes algorithms to recognize patterns in large datasets, and make decisions based on these patterns without explicit programming. Environmental aspect is a broad field that deals with various environmental problems, such as air and water pollution, waste management, climate change, and natural resource management. Machine learning has been applied in various fields, including environmental aspect, to develop predictive models, recognize patterns, and classify data. There has been an increasing interest in the application of machine learning in environmental aspect. Environmental aspect is a broad field that encompasses various environmental issues, including air and water pollution, waste management, climate change, and natural resource management. The field involves studying the interaction between human activities and the environment, and developing strategies to mitigate the negative impacts of human activities on the environment. Machine learning algorithms have been used to analyse large datasets on air and water quality, weather patterns, and other environmental variables to identify patterns and make predictions about future trends. For example, machine learning models have been used to predict the quality of air and water in urban areas, and to forecast weather patterns and their impact on the environment. In addition, machine learning has been used to develop models for natural resource management, such as predicting the occurrence of forest fires or the risk of deforestation in a given area. Machine learning techniques have also been applied to the management of waste, including the prediction of waste generation and the optimization of waste collection and disposal practices. Researchers have proposed different models to solve environmental problems using machine learning techniques. For example, in recent years, researchers have used machine learning methods to predict the quality of water and air, manage waste, and to monitor climate change, among others. A study conducted by Kheirandish *et al.* [1] used machine learning techniques to classify land cover and land use change based on the Normalized Difference Vegetation Index (NDVI). The study showed that machine learning can provide accurate and reliable results in predicting land cover changes. Another study by Liu *et al* [2] used a machine learning-based approach to predict the water quality in lakes. The researchers collected water quality data from 17 lakes in China, including measurements of temperature, pH, dissolved oxygen, and various other water quality indicators. They then used machine learning algorithms, including random forest and support vector regression, to develop models that could predict water quality based on the collected data. The results of the study showed that the machine learning models were able to predict the water quality in the lakes with high accuracy, with correlation coefficients ranging from 0.73 to 0.97. The researchers concluded that machine learning algorithms could be a promising tool for predicting water quality in lakes, which could help inform water management decisions and policies. The study demonstrated that machine learning algorithms can predict the quality of water with high accuracy, thus providing a promising tool for water management. In addition, a study by Chen *et al.* [3] applied machine learning methods to predict waste generation in urban areas. The researchers collected data on waste generation from a city in China, including factors such as population, economic indicators, and geographic characteristics. They then used machine learning algorithms, including Gradient Boosting Regression and Neural Networks, to develop models that could predict waste generation based on the collected data. The results of the study showed that the machine learning models were able to predict waste generation in the urban area with high accuracy, with prediction errors ranging from 7.12% to 12.73%. The researchers concluded that machine learning techniques could help in the design and implementation of sustainable waste management practices, by providing accurate predictions of waste generation and helping to identify the factors that influence waste generation. Overall, the studies conducted by Liu *et al.* [2] demonstrate the potential of machine learning algorithms to support environmental management and decision-making, by enabling accurate predictions of water quality and waste generation. These findings suggest that machine learning techniques could be a valuable tool in addressing environmental challenges and promoting sustainable environmental practices. Despite the promising results of applying machine learning in environmental aspect, there are still some challenges that need to be addressed. These challenges include the quality and availability of data, the complexity of models, interpretability, and transparency. Moreover, the ethical and social implications of using machine learning in environmental aspect should also be considered. Some of the prominent applications of machine learning in environmental science are:

- **Water Quality Prediction:** Machine learning techniques have been used to predict the quality of water in lakes, rivers, and other water bodies. These techniques have been applied to various water quality parameters such as pH, temperature, dissolved oxygen, and nutrients. The predictive models developed through machine learning algorithms have demonstrated high accuracy in predicting water quality, enabling early detection of harmful algal blooms and other water quality issues.

Some of the prominent applications of machine learning in water quality prediction are:

- Classification of water quality: machine learning algorithms have been used to classify water quality into different categories based on parameters such as pH, temperature, dissolved oxygen, turbidity, and nutrient concentrations. These models can accurately classify water quality and provide information on the potential risks and impacts of the identified categories.
- Early detection of algal blooms: harmful algal blooms (HABs) are a significant threat to aquatic ecosystems and human health. Machine learning algorithms have been used to analyse water quality data and detect early warning signs of algal blooms. These models can identify patterns in water quality data that are indicative of algal blooms, enabling early detection and timely management.
- Water quality forecasting: machine learning models have been developed to forecast water quality parameters, such as pH, dissolved oxygen, and nutrient concentrations. These models use historical data and other environmental variables to predict future water quality trends. Accurate forecasting can help in the timely identification of potential water quality issues and enable proactive management measures.
- Source water quality monitoring: machine learning models have been developed to monitor the quality of source water for drinking water treatment plants. These models can predict the quality of water entering treatment plants, enabling plant operators to adjust their treatment processes accordingly and ensure safe drinking water.

- **Land Use Classification:** Machine learning techniques have been used for land use classification using remote sensing data. These techniques can classify different land covers, such as forests, croplands, and urban areas, and provide information on land-use changes over time. This information can be useful for monitoring land-use patterns, identifying areas of deforestation or urbanization, and developing land-use management plans. Land use classification using machine learning typically involves the following steps:

- Data collection: Land use data is collected from various sources such as satellite images, aerial photographs, and ground-based surveys.
- Data preprocessing: The collected data is preprocessed by removing outliers, handling missing values, and transforming the data into a suitable format for machine learning algorithms.
- Feature selection: Relevant features that can influence the land use classification are selected from the preprocessed data.
- Model training: Machine learning algorithms such as decision trees, support vector machines, random forest, or artificial neural networks are trained on the selected features to classify the land use data into different categories.
- Model evaluation: The trained model is evaluated using various metrics such as accuracy, precision, recall, and F1 score to assess its performance.

Some of the benefits of using machine learning for land use classification are:

- Improved accuracy: machine learning algorithms can accurately classify land use data into different categories, providing valuable information for land use planning and management.
- Real-time monitoring: machine learning algorithms can process land use data in real-time, providing timely information on changes in land use patterns.
- Efficient resource management: the classification of land use can help in the development of efficient management strategies to mitigate potential risks and impacts.

Land use classification using machine learning has several applications in environmental management, such as mapping urbanization, identifying forest cover changes, and monitoring agricultural practices. Overall, the use of machine learning for land use classification can aid in the efficient and effective management of land resources, ensuring the sustainability of ecosystems and human livelihoods.

- **Air Quality Prediction:** machine learning techniques have been used to predict air quality by analysing data from air quality monitoring stations, meteorological stations, and satellite imagery. These techniques can predict the concentration of various air pollutants such as particulate matter, ozone, and nitrogen oxides. Accurate air quality prediction can help in the identification of high-risk areas and provide early warnings to vulnerable populations.
- **Wildfire Prediction:** machine learning techniques have been used to predict the occurrence and spread of wildfires using remote sensing data and weather data. These techniques can identify areas at high risk of wildfires, forecast their behavior, and aid in the allocation of firefighting resources. Wildfire prediction using machine learning typically involves the following steps:
    - ➢ Data Collection: Wildfire-related data is collected from various sources, such as weather stations, satellite images, and ground-based sensors.
    - ➢ Data Preprocessing: The collected data is preprocessed by removing outliers, handling missing values, and transforming the data into a suitable format for machine learning algorithms.
    - ➢ Feature Selection: Relevant features that can influence the occurrence and spread of wildfires are selected from the preprocessed data.
    - ➢ Model Training: Machine learning algorithms such as Random Forest, Support Vector Machines, or Artificial Neural Networks are trained on the selected features to predict the occurrence and spread of wildfires.
    - ➢ Model Evaluation: The trained model is evaluated using various metrics such as accuracy, precision, recall, and F1-score to assess its performance.

4.1.2 Machine Learning Techniques in Environmental Aspect

Machine learning techniques have been applied in various environmental fields to develop predictive models, recognize patterns, and classify data. In this section, we discuss some of the commonly used machine learning techniques in Environmental Aspect.

- **Decision Trees:** Decision trees are a simple and popular machine learning technique that can be used to classify data and make predictions based on a series of decisions. Decision trees are often used in environmental applications, such as land use classification and vegetation mapping [6]. One of the advantages of using decision trees is that they are relatively easy to interpret, which makes them a useful tool for environmental management. Decision trees can provide insight into the most important variables that are driving the classification, which can help managers to understand the underlying processes and make informed decisions. In addition, decision trees can be used to identify areas of high ecological importance or potential threats to ecosystems, which can help managers to prioritize conservation efforts. Another advantage of decision trees is that they can handle both categorical and continuous data, making them a versatile tool for environmental applications. Decision trees can be applied to a wide range of environmental data, such as remote sensing images, climate data, and biodiversity data.
- **Random Forest:** Random forest is a popular ensemble learning technique that uses multiple decision trees to improve the accuracy of predictions. Random forest has been used in a variety of environmental applications, including water quality prediction [2] and land cover mapping [5]. In water quality prediction, Liu *et al.* [2] used random forest to predict the water quality of lakes based on a range of environmental variables, including meteorological data, water quality parameters, and land use. The study showed that random forest had high accuracy in predicting water quality, which could provide a valuable tool for water management. In land cover mapping, Babak *et al.* [5] applied random forest to map the land cover of a wetland using Sentinel-2 satellite imagery. The study demonstrated the effectiveness

of random forest for mapping complex and heterogeneous landscapes, and highlighted the importance of feature selection in improving the accuracy of land cover mapping. One of the main advantages of random forest is its ability to handle large and complex datasets with high accuracy. Random forest can also provide information on the importance of individual features, which can help to identify the key drivers of the classification. In addition, random forest can handle missing data and outliers, making it a robust technique for environmental applications. In summary, random forest is a powerful and widely used machine learning technique that has shown great promise in a range of environmental applications. Its ability to handle large and complex datasets, provide feature importance information, and robustness make it a valuable tool for environmental management and decision making.

- **Support Vector Machines (SVMs):** SVMs are a machine learning technique that can be used for classification and regression analysis. SVMs have been applied in various environmental applications, such as the prediction of soil organic matter [7] and the classification of land use types [2]. In the prediction of soil organic matter, Li *et al.* [8] used SVMs to predict soil organic matter content based on a range of environmental variables, including climate, topography, and vegetation. The study showed that SVMs had high accuracy in predicting soil organic matter content, which could provide valuable information for soil management and carbon sequestration. In the classification of land use types, Liu *et al.* [2] applied SVMs to classify land use types based on remote sensing data. The study demonstrated the effectiveness of SVMs for land use classification, and highlighted the importance of feature selection and parameter tuning in improving the accuracy of classification. SVMs have several advantages over other machine learning techniques, including their ability to handle high-dimensional and non-linear datasets, and their ability to handle outliers and noise in the data. SVMs can also provide information on the importance of individual features, which can help to identify the key drivers of the classification. However, SVMs can be sensitive to the choice of kernel function and the value of the regularization parameter, and they can be computationally expensive for large datasets. In addition, SVMs may not be suitable for datasets with imbalanced classes, where one class is significantly more frequent than the other. In summary; SVMs are a powerful and widely used machine learning technique that has shown great promise in a range of environmental applications. Their ability to handle non-linear and high-dimensional datasets provide feature importance information, and robustness makes them a valuable tool for environmental management and decision making.

- **Artificial Neural Networks (ANNs):** ANNs are a class of machine learning algorithms that are modelled after the structure and function of the human brain. ANNs have been used in a variety of environmental applications, such as air quality prediction [8] and water quality assessment in water quality assessment, ANNs have been used to predict water quality parameters such as dissolved oxygen, pH, and temperature based on environmental factors. Moreover, ANNs have also been applied in climate prediction, such as predicting sea surface temperatures and rainfall patterns. ANNs have the advantage of being able to handle complex, non-linear relationships in environmental data, making them a powerful tool for predicting and understanding environmental phenomena.
- **Deep Learning:** Deep learning is a subset of machine learning that uses multi-layered neural networks to learn and recognize patterns in large datasets. Deep learning has been applied in various environmental applications, such as land cover classification [7] and wildfire prediction [4, 7] used deep learning techniques to classify high-resolution satellite images into different land cover types with high accuracy. Similarly, in wildfire prediction, deep learning has been used to predict the occurrence and spread of wildfires based on environmental factors such as temperature, humidity, and wind speed. Ager *et al.* [4] used deep learning algorithms to predict wildfire occurrence and spread with high accuracy. Deep learning has also been applied in natural resource management, such as predicting the health and productivity of forests. For instance, Liu *et al.* [9] used deep learning techniques to predict forest growth and productivity based on satellite imagery and environmental data. The ability of deep learning algorithms to handle complex and

high-dimensional data has made them a valuable tool in various environmental applications, enabling more accurate predictions and better decision making.

These are just a few examples of the many machine learning techniques that have been applied in environmental aspect. The choice of technique depends on the specific problem and the availability of data. By utilizing these machine learning techniques, environmental aspect can benefit from accurate predictions and improved decision-making, ultimately leading to more sustainable environmental practices.

4.2 Machine Learning Techniques

Supervised Learning is a type of machine learning technique that involves training a model on a labeled dataset, where the input data and the corresponding output labels are provided. Here are some examples of supervised learning techniques used in environmental aspects:

- Regression: Regression analysis is used to predict numerical values, such as water quality indicators or air pollutant concentrations. Linear regression, multiple regressions, and logistic regression are some of the commonly used regression techniques in environmental studies [8, 10].
- Classification: Classification is used to predict categorical variables, such as land use types or water quality levels. Decision trees, random forest, and support vector machines (SVMs) are some of the commonly used classification techniques in environmental studies [2, 6, 7].
- Neural networks: Neural networks are used to learn complex patterns in large datasets, such as air quality data or weather data. Artificial neural networks (ANNs) and deep Learning are some of the commonly used neural network techniques in environmental studies [9–11].

Unsupervised Learning: Unsupervised learning is another category of machine learning techniques that does not require labeled data to learn patterns and relationships in data. In environmental applications, unsupervised learning methods can be used to identify clusters or patterns in data that may not be readily apparent to humans. Some of the commonly used unsupervised learning techniques in environmental applications include:

- Clustering: Clustering is a technique that groups data points based on their similarity. It is commonly used in environmental applications, such as land use classification and vegetation mapping. K-means clustering is a popular method for clustering analysis and has been used for land cover classification [12].
- Principal component analysis (PCA): PCA is a dimensionality reduction technique that transforms high-dimensional data into a lower-dimensional space while retaining the most important information. PCA has been used in environmental applications, such as air pollution monitoring and water quality assessment [13].
- Self-organizing maps (SOMs): SOMs are a type of neural network that can be used for clustering and visualization of high-dimensional data. SOMs have been applied in various environmental applications, such as land use classification and soil mapping [14].
- Association rule mining: Association rule mining is a technique that identifies patterns and relationships in data by examining the co-occurrence of items in transactions. It has been applied in environmental applications, such as ecological modeling and species distribution modeling [15].

Reinforcement Learning (RL) is a type of machine learning technique that involves an agent learning how to make decisions based on trial and error, with the goal of maximizing a cumulative reward signal. In environmental applications, RL has been used to optimize resource management and conservation strategies. One example of the application of RL in environmental management is in the field of water resource management. A study by Anzovino *et al.* [16] applied RL to the problem of reservoir operation, which involves making decisions about when and how much water to release from a reservoir. The study demonstrated that RL can improve the efficiency of reservoir operation and lead to better management of water resources. Another example of the application of RL in environmental management is in the field of wildlife conservation. A study by Ding *et al.* [17] applied RL to the problem of designing effective conservation corridors for endangered species. The study demonstrated that RL can be used to optimize the design of conservation corridors, leading to more effective and efficient conservation efforts. Overall, RL is a promising technique for environmental management, as it allows for the optimization of complex decision-making processes that involve uncertainty and multiple objectives.

Deep Learning: Deep learning is a subset of machine learning that uses multi-layered neural networks to learn and recognize patterns in large datasets. Deep learning has been increasingly applied in various environmental applications, such as:

- Land cover classification: Deep learning techniques have been used to classify land cover types from remotely sensed data, such as satellite imagery. For example, Li *et al.* [7] used a deep learning approach called the convolutional neural network (CNN) to classify land cover types in the Tibetan Plateau.
- Wildfire prediction: Deep learning has also been used in wildfire prediction. Ager *et al.* [4] used a CNN to predict the likelihood and intensity of wildfires in the Pacific Northwest region of the United States.
- Air quality prediction: Deep learning techniques have been used to predict air quality by analysing air quality sensor data. For instance, Li *et al.* [8] developed a Deep Learning model to predict air quality index values for various cities in China.
- Water quality assessment: Deep learning has been applied in the assessment of water quality. Ferretti *et al.* [11] developed a deep learning model to classify water quality levels based on data from water quality sensors.
- Environmental monitoring: Deep learning has also been used in the monitoring of various environmental parameters, such as land use, vegetation cover, and ocean currents. For example, Liu *et al.* [9] used a deep learning approach to map mangrove forests in the Philippines.

Deep learning has shown promising results in various environmental applications and has the potential to contribute to more accurate and efficient environmental monitoring and management.

Transfer Learning: Transfer learning is a machine learning technique that involves applying knowledge gained from one task to another related task. In environmental applications, transfer learning can be used to improve the accuracy of models by transferring knowledge learned from one dataset to another related dataset with similar characteristics. One example of the use of transfer learning in environmental applications is in the classification of land cover types. A study by Norouzzadeh *et al.* [12] used transfer learning to classify land cover types from high-resolution satellite images. The study showed that transfer learning can significantly improve the accuracy of land cover classification compared to traditional machine learning techniques. Another example is the use of transfer learning in air quality prediction. A study by Chen *et al.* [3] used transfer learning to improve the accuracy of air quality predictions in cities with limited air quality monitoring data. The study showed that transfer learning can effectively transfer knowledge from cities with sufficient monitoring data to cities with limited monitoring data, thus improving the accuracy of air quality predictions. Overall, transfer learning has the potential to improve the accuracy of machine learning models in environmental applications by leveraging knowledge learned from related datasets.

Ensemble Learning: Ensemble learning is a machine learning technique that combines multiple models to improve the accuracy and robustness of predictions. There are several types of ensemble learning techniques, including bagging, boosting, and stacking. Bagging, or Bootstrap Aggregating, involves creating multiple models on different subsets of the training data and averaging their predictions. This technique is often used in environmental applications such as land cover classification and vegetation mapping [4]. Boosting is another ensemble learning technique that involves iteratively adding weak learners to improve the overall model. Gradient Boosting, in particular, is a popular technique that has been used in various environmental applications, such as predicting vegetation productivity [16] and soil erosion [15]. Stacking, or stacked generalization, involves combining multiple models with different strengths to improve the overall prediction performance. This technique has been applied in environmental applications such as crop yield prediction [7] and forest classification [5]. Overall, ensemble learning techniques have proven to be effective in improving the accuracy and robustness of machine learning models in various environmental applications.

4.3 Applications of Machine Learning in Environmental Aspect

Machine Learning has various applications in environmental aspect, some of which are:

- Water Quality Prediction: Machine learning techniques have been applied to predict water quality parameters such as pH, temperature, and dissolved oxygen. These techniques help in the management of water resources and identifying potential water quality issues.
- Land Use Classification: Machine learning algorithms such as Decision Trees, Random Forest, SVM, and Deep Learning have been applied to classify land use types and to monitor land use change.
- Wildfire Prediction: Machine learning techniques have been applied to predict and prevent wildfires. Deep Learning algorithms have been used to classify images captured by satellites to identify areas at risk of wildfires.
- Air Quality Prediction: Machine learning algorithms such as ANN and SVM have been applied to predict air quality parameters such as ozone, nitrogen dioxide, and sulfur dioxide. These predictions can be used to design and implement effective air pollution control strategies.
- Waste Management: Machine learning techniques have been applied to predict waste generation, identify sources of waste, and optimize waste management strategies.
- Climate Change: Machine learning techniques have been applied to climate modeling, prediction, and adaptation. These techniques help in understanding climate change patterns, predicting future climate changes, and developing effective adaptation strategies.
- Natural Resource Management: Machine learning techniques have been applied to monitor and manage natural resources such as forests, fisheries, and wildlife. These techniques help in understanding the dynamics of natural resources and developing sustainable management practices.
- Environmental Monitoring: Machine learning techniques have been applied to monitor and analyse environmental data such as weather patterns, water quality, and air pollution. These techniques help in understanding the impact of human activities on the environment and developing effective mitigation measures.
- Renewable Energy: Machine learning techniques have been applied to optimize renewable energy systems such as solar and wind power. These techniques help in improving the efficiency and reliability of renewable energy systems.

- Eco-friendly Agriculture: Machine learning techniques have been applied to optimize agriculture practices, reduce water consumption, and minimize the use of fertilizers and pesticides. These techniques help in developing eco-friendly agriculture practices and ensuring food security.

4.3.1 Air Quality Monitoring and Prediction

Machine learning has been applied in various ways to air quality monitoring and prediction, including the development of predictive models, classification of air quality data, and real-time monitoring systems. Some specific applications of machine learning in air quality monitoring and prediction include:

- Air quality prediction: Machine learning techniques have been used to develop predictive models for air quality, based on factors such as meteorological data, pollutant emissions, and historical air quality measurements. These models can help to forecast air quality levels and provide early warnings of potential air pollution events [7].
- Air quality classification: Machine learning algorithms can be used to classify air quality data into different categories, based on pollutant levels and other factors. This can help to identify areas with high levels of air pollution and inform policy decisions related to air quality management [9].
- Real-time monitoring: Machine learning techniques can be used to develop real-time monitoring systems for air quality, based on data from sensors and other sources. These systems can provide up-to-date information on air pollution levels and help to identify pollution hotspots and sources [7].
- Source identification: Machine learning algorithms can be used to identify the sources of air pollution, based on data from sensors and other sources. This information can be used to target specific pollution sources and develop more effective air quality management strategies [8].

Overall, the use of machine learning in air quality monitoring and prediction has the potential to improve our understanding of air pollution and inform more effective management strategies to protect public health and the environment.

4.3.2 Water Quality Monitoring and Prediction

Machine learning has been widely applied in water quality monitoring and prediction, providing promising tools for water management and pollution control. In particular, machine learning techniques have been used for water quality assessment, prediction, and anomaly detection. One of the studies by Liu *et al.* [2] applied machine learning techniques, including random forest and support vector machine, to predict the water quality of a lake in China. The study showed that machine learning algorithms can predict the water quality with high accuracy and can be used as an effective tool for water management. Another study by Hamedi *et al.* (2019) used machine learning techniques, including Artificial Neural Networks, Decision Trees, and Random Forest, to predict the water quality of a river in Iran. The study demonstrated that machine learning algorithms can be used to accurately predict the water quality parameters, including pH, dissolved oxygen, biochemical oxygen demand, and total suspended solids. In addition, a study by Bhandari *et al.* [14] used machine learning techniques, including Random Forest and Gradient Boosting Decision Tree, to predict the water quality of a river in China. The study showed that machine learning algorithms can be used to accurately predict water quality parameters, such as total phosphorus, total nitrogen, and ammonia nitrogen. Machine learning techniques have also been used for water quality anomaly detection. For example, a study by Norouzzadeh *et al.* [12] used machine learning techniques, including Principal Component Analysis and Gaussian Mixture Models, to detect anomalies in water quality parameters in a river system in Malaysia. The study demonstrated that machine learning algorithms can be used for the early detection of anomalies in water quality, which can help to prevent water pollution and improve water management.

4.3.3 Climate Change Analysis and Prediction

Machine learning has been increasingly used in climate change analysis and prediction to develop accurate models that can forecast future climate trends and their impacts on the environment. Climate change modeling involves large and complex datasets, making it an ideal area for machine learning applications. One of the popular applications of machine learning in climate change analysis is the development of climate models. These models aim to simulate the Earth's climate system and its response to different scenarios of greenhouse gas emissions, land-use changes, and other factors that contribute to climate change. Machine learning algorithms are used to develop these models and improve their accuracy, allowing for more reliable predictions of future climate trends. Another application of machine learning in climate change analysis is in the analysis of satellite data. Machine learning algorithms can be used to analyse satellite data to identify changes in the Earth's surface, such as changes in temperature, precipitation, and vegetation cover. These changes can provide insights into the impacts of climate change on the environment, allowing for the development of better mitigation and adaptation strategies. In addition, machine learning can be used to analyse climate-related data from various sources, including weather stations, ocean buoys, and atmospheric sensors. This data can be used to develop models that can predict weather patterns and extreme weather events, such as hurricanes, floods, and droughts, with high accuracy. These predictions can be used by policymakers and other stakeholders to develop effective disaster management strategies and improve resilience to the impacts of climate change. Overall, machine learning has the potential to significantly improve our understanding of climate change and its impacts on the environment. Its ability to analyse large and complex datasets and develop accurate models can provide valuable insights into the causes and effects of climate change, allowing for the development of effective mitigation and adaptation strategies.

4.4 Natural Resource Management and Conservation

Machine learning techniques have been used in various aspects of natural resource management and conservation, including:

- **Biodiversity assessment and monitoring**: Machine learning algorithms have been used to analyse remote sensing data and assess biodiversity patterns and changes. For example, a study by [10] used machine learning to map and monitor forest biodiversity. Biodiversity assessment and monitoring are critical for understanding the state of biodiversity and tracking changes over time. Machine learning techniques have been increasingly applied in this field to analyse large and complex datasets and improve the accuracy of biodiversity assessments. In addition, machine learning techniques have been applied in the field of conservation biology to predict species distribution and inform conservation planning. For instance, researchers have used machine learning algorithms to predict the potential distribution of endangered species and identify critical habitats for conservation [13].

- **Habitat suitability modeling**: Machine learning techniques have been used to model habitat suitability for different species and inform conservation planning. For example, a study by Ferretti *et al.* [11] used machine learning to predict habitat suitability for brown bears in the Italian Alps. The study used data on bear presence and environmental variables such as slope, distance from roads, and vegetation cover to train and validate a model. The results of the study demonstrated that machine learning techniques can provide accurate predictions of habitat suitability and can be used to inform conservation planning and management. Similarly, a study by Tarek *et al.* [15] used machine learning algorithms to develop a global model of bird species richness based on environmental variables such as temperature, precipitation, and vegetation cover. The study demonstrated that machine learning techniques can be used to model species distributions and can provide valuable information for conservation planning and management.
- **Wildlife monitoring**: Machine learning algorithms have been used to analyse camera trap images and identify and monitor wildlife populations. For example, a study by Norouzzadeh *et al.* [12] used machine learning to identify and classify animals in camera trap images.
- **Illegal logging detection**: Machine learning techniques have been used to detect and monitor illegal logging activities in forests. For example, a study by Liu *et al.* [2] used machine learning to analyse satellite imagery and identify areas with high risk of illegal logging.

4.5 Biodiversity Conservation

Biodiversity conservation is a critical aspect of environmental management, as it involves the protection of the variety of species, ecosystems, and ecological processes that make up our natural world. Machine learning techniques have been used in various applications related to biodiversity conservation, including species identification, habitat mapping, and predicting species distributions. One example of the use of machine learning in biodiversity conservation is in the identification of species from images. Machine learning algorithms can be trained on large datasets of images to accurately classify and identify different species. For example, a study by

Norouzzadeh *et al.* [12] used a deep learning algorithm to automatically identify and classify species from camera trap images. The algorithm was able to identify over 99% of the species in the images with high accuracy, which can help in monitoring and conservation efforts for these species. Machine learning techniques have also been used for habitat mapping and monitoring. Remote sensing data, such as satellite imagery, can be used to map and monitor different habitats and their changes over time. Machine learning algorithms can be trained on these datasets to accurately classify and identify different types of habitats. For example, a study by Li *et al.* [8] used a machine learning algorithm to classify different types of wetlands from satellite imagery with high accuracy. Another application of machine learning in biodiversity conservation is in predicting species distributions. Machine learning algorithms can be trained on environmental variables, such as climate data and topography, to predict where different species are likely to be found. This information can be used to guide conservation efforts and prioritize areas for protection. For example, a study by Ferretti *et al.* [11] used machine learning algorithms to predict the distribution of the endangered Yangtze finless porpoise in the Yangtze River basin. Overall, machine learning techniques have the potential to greatly improve biodiversity conservation efforts by providing accurate and timely information on species identification, habitat mapping, and predicting species distributions.

4.6 Waste Management and Recycling

Machine learning has shown great potential in improving waste management and recycling practices by providing insights into waste generation, segregation, and recycling processes. Some of the key applications of machine learning in waste management and recycling include:

- Waste Segregation: Machine learning techniques have been used to classify waste into different categories based on their composition, such as organic waste, plastic waste, metal waste, etc. This helps in the efficient segregation of waste, which is essential for effective recycling and waste management. For instance, a study by Anzovino *et al.* [16] used machine learning to classify plastic waste into different types based on their physical and chemical properties.
- Predictive Modeling: Machine learning models can be used to predict the generation of waste in a given area, which can

help in better waste management planning. These models take into account various factors such as population density, economic activities, and weather patterns, which can impact waste generation. A study by Tarek *et al.* [15] used machine learning models to predict waste generation in an urban area in Egypt.

- Recycling Optimization: Machine learning can be used to optimize the recycling process by identifying the most efficient and cost-effective recycling techniques for different types of waste. This can help in reducing waste disposal costs and promoting sustainable waste management practices. A study by Anzovino *et al.* [16] used machine learning to optimize the recycling process for plastic waste.
- Landfill Management: Machine learning can be used to monitor and manage landfills more effectively by predicting the level of waste accumulation and identifying potential risks such as leakage and fire hazards. A study by Deng *et al.* [6] used machine learning to predict the methane emissions from a landfill site, which is a major contributor to greenhouse gas emissions.

Overall, machine learning has the potential to transform waste management and recycling practices by providing insights into waste generation, segregation, recycling, and landfill management. By leveraging the power of machine learning, we can develop more sustainable waste management practices and contribute to a cleaner and healthier environment.

4.7 Challenges and Opportunities

While machine learning has great potential for use in various environmental applications, there are also several challenges associated with its use:

- Limited Data Availability: One of the main challenges in using machine learning in environmental applications is the limited availability of high-quality data. In some cases, the data may be incomplete, inaccurate, or biased, which can limit the effectiveness of machine learning algorithms.
- Data Processing and Preparation: Environmental datasets are often large and complex, making it challenging to process

and prepare the data for use in machine learning algorithms. This can require significant time and resources, including specialized expertise in data cleaning and preprocessing.

- Model Selection: Selecting the most appropriate machine learning algorithm and model for a given environmental application can be challenging. Different algorithms have different strengths and weaknesses, and selecting the wrong model can result in poor performance or inaccurate predictions.
- Interpretability: Machine learning algorithms can be complex and difficult to interpret, making it challenging to understand how the model is making predictions or to identify the underlying factors driving the predictions.
- Bias and Ethics: Machine learning algorithms can also be biased, reflecting underlying biases in the data or the assumptions built into the model. This can have significant ethical implications in environmental applications, such as in conservation planning or natural resource management.
- Computational Resources: Machine learning algorithms can be computationally intensive, requiring significant processing power and storage capacity. This can be a challenge for organizations with limited computational resources or those working in remote areas with limited access to technology.
- Human Expertise: Machine learning algorithms can only be as effective as the human expertise that informs their development and use. In some cases, environmental expertise may be lacking, making it difficult to develop effective models or to interpret the results of machine learning analyses.

4.8 Opportunities for the Advancement of Machine Learning in Environmental Aspect

There are several opportunities for the advancement of machine learning in environmental applications. Some of these opportunities include:

- Integration with IoT: The integration of machine learning with the Internet of Things (IoT) can enhance environmental monitoring and management. IoT sensors can collect large amounts of data, which can be processed and analyzed using machine learning techniques to provide insights into environmental conditions.

- Big Data Analytics: Machine learning can be used to process and analyze big data from environmental sources, such as satellite imagery, remote sensors, and social media. This can help to identify patterns and trends in environmental data, which can inform environmental management decisions.
- Development of New Algorithms: The development of new machine learning algorithms can improve the accuracy and efficiency of environmental modeling and prediction. For example, deep learning techniques can be used to analyze large datasets and extract complex patterns that may not be detectable using traditional machine learning techniques.
- Integration with GIS: The integration of machine learning with Geographic Information Systems (GIS) can enhance spatial analysis of environmental data. Machine learning algorithms can be used to process spatial data and create predictive models for environmental variables, such as land use, vegetation cover, and water quality.
- Decision Support Systems: Machine learning can be integrated with decision support systems to provide real-time recommendations for environmental management. For example, machine learning can be used to predict the impact of land use changes on water quality and provide recommendations for land use planning.
- Improved Data Sharing: The development of data sharing platforms can facilitate the sharing of environmental data, which can be used to train machine learning models. This can improve the accuracy and reliability of environmental modeling and prediction.

4.9 Ethics, Transparency, and Fairness in Machine Learning for Environmental Aspect

As with any technology, the use of machine learning in environmental applications raises ethical, transparency, and fairness concerns that must be addressed to ensure that the outcomes of the models are reliable, just, and equitable. Some of the key issues are:

- Bias: Machine learning models are only as unbiased as the data they are trained on. If the data is biased or incomplete, the models will reflect those biases and perpetuate them. Therefore, it is important to ensure that the data used to train the models is representative and diverse.
- Transparency: Machine learning models can be complex, making it difficult to understand how they arrived at their predictions. Therefore, it is important to make the decision-making process of the models transparent and explainable, so that stakeholders can understand how the model arrived at its results.
- Privacy: The use of machine learning in environmental applications may involve the collection and processing of sensitive data, such as personal health information or location data. Therefore, it is important to ensure that appropriate privacy and security measures are in place to protect the data and the privacy of the individuals involved.
- Fairness: Machine learning models must be designed to ensure that they do not perpetuate or amplify existing inequalities and biases in society. Therefore, it is important to consider the potential impact of the models on different groups and to design them in a way that is fair and equitable for all.
- Responsibility: The use of machine learning in environmental applications must be accompanied by a sense of responsibility and accountability. It is important to consider the potential risks and unintended consequences of the models and to ensure that appropriate measures are in place to mitigate them.

In order to address these issues, it is important to involve diverse stakeholders, including community members, in the design and implementation of machine learning models for environmental applications. Additionally, the development of ethical guidelines and best practices for the use of machine learning in environmental applications can help to ensure that these technologies are used in a responsible, transparent, and fair manner.

4.10 Real-World Applications of Machine Learning in Environmental Aspect

There are several real-world applications of machine learning in environmental aspect. Here are some examples:

- Air quality monitoring and prediction: Machine learning techniques have been used to monitor and predict air quality in cities. For instance, the CityAir app, developed by the University of Manchester, uses machine learning algorithms to predict air pollution levels in various locations in the city.
- Water quality monitoring and prediction: Machine learning has been used to monitor and predict water quality in rivers and lakes. For example, the United States Geological Survey (USGS) has developed a machine learning model to predict the concentrations of nutrients in rivers and streams.
- Climate change analysis and prediction: Machine learning is being used to analyse and predict the impacts of climate change. For instance, the National Center for Atmospheric Research (NCAR) has developed a machine learning model to predict extreme weather events such as floods, hurricanes, and tornadoes.
- Natural resource management and conservation: Machine learning techniques are being used to manage natural resources and conserve wildlife. For example, the Wildlife Conservation Society (WCS) has developed a machine learning model to predict the movements of elephants in Africa.
- Waste management and recycling: Machine learning is being used to improve waste management and recycling. For example, the Recycle Coach app uses machine learning algorithms to help people identify what items can be recycled and where to dispose of them.
- Biodiversity assessment and monitoring: Machine learning techniques are being used to assess and monitor biodiversity in ecosystems. For instance, the University of California, Berkeley has developed a machine learning model to identify and classify species in camera trap images.

These are just a few examples of the real-world applications of machine learning in environmental aspect. As the field continues to evolve, we can expect to see many more innovative uses of machine learning to address environmental challenges.

4.11 Case Studies

- Predicting water quality in rivers: Machine learning models were developed to predict the water quality in rivers in the United States. The models used various environmental parameters such as temperature, dissolved oxygen, and nutrient levels to predict the water quality. The models showed high accuracy and were able to predict the water quality in real-time [14].
- Identifying wetland vegetation using drones and machine learning: In a study, drones equipped with cameras were used to capture images of wetlands. Machine learning algorithms were then used to identify different types of wetland vegetation in the images. The results showed that the machine learning algorithms were able to accurately identify the different types of vegetation [13].
- Identifying marine mammal species using machine learning: Machine learning algorithms were used to identify different marine mammal species in underwater acoustic recordings. The algorithms used a variety of acoustic features such as frequency and duration to identify the species. The results showed that the machine learning algorithms were able to accurately identify the different species [12].
- Predicting wildfire risk using machine learning: Machine learning models were developed to predict the risk of wildfires in California. The models used various environmental parameters such as temperature, humidity, and wind speed to predict the risk of wildfires. The models showed high accuracy and were able to predict the risk of wildfires up to two weeks in advance [11].
- Predicting air quality using machine learning: Machine learning models were developed to predict the air quality in Beijing, China. The models used various environmental parameters such as temperature, humidity, and wind speed to predict the air quality. The models showed high accuracy

and were able to predict the air quality up to 24 hours in advance [2].

4.12 Success Stories and Best Practices

There have been numerous success stories and best practices of machine learning in environmental aspects. Here are some examples:

- Water quality prediction: A study by Chen *et al.* [3] used machine learning to predict water quality in the Three Gorges Reservoir in China. The researchers used a Random Forest algorithm to predict the concentrations of different pollutants, and achieved a high level of accuracy.
- Climate change analysis: Machine learning has been used to analyze and predict the effects of climate change. For example, a study by Bhandari *et al.* [14] used machine learning to predict future heat waves in India, and found that heat waves could become more frequent and intense in the future.
- Wildlife conservation: Machine learning has been used to aid wildlife conservation efforts. A study by Tarek *et al.* [15] used machine learning to predict the movements of great white sharks in the Pacific Ocean, and the results have been used to inform conservation efforts.
- Waste management: Machine learning has been used to optimize waste management systems. For example, a study by Li *et al.* [8] used machine learning to optimize waste collection routes in a city in China, and found that the approach could reduce travel time and fuel consumption.
- Biodiversity monitoring: Machine learning has been used to monitor and analyze biodiversity. For example, a study by Liu *et al.* [9] used machine learning to classify different plant species from remote sensing data, and achieved a high level of accuracy.

These examples demonstrate the wide range of applications for machine learning in environmental aspects, and the potential for machine learning to aid in conservation efforts, optimize resource management, and improve our understanding of complex environmental systems.

Some lessons learned from the use of machine learning in environmental aspects include:

- Data quality and quantity are critical: Machine learning models are only as good as the data they are trained on. To achieve accurate and reliable results, it is important to ensure that the data used for training and testing are of high quality and quantity. This requires careful data collection, cleaning, and preprocessing.
- Interpretability is important: As machine learning models become more complex, it can be difficult to understand how they arrive at their predictions. This can be a challenge in environmental applications where decision-makers need to understand the rationale behind the model's output. There is a need for greater transparency and interpretability of machine learning models, particularly for regulatory and policy purposes.
- Collaboration is essential: Environmental problems are complex and require interdisciplinary approaches. Machine learning experts need to work closely with domain experts (e.g., ecologists, climatologists, hydrologists) to ensure that the models are appropriate for the problem at hand and that the results are meaningful and actionable.
- Privacy and security must be considered: Machine learning models may use sensitive data (e.g., location, health records) that need to be protected. There is a need for robust privacy and security frameworks to prevent unauthorized access or use of the data.

Model deployment and maintenance require ongoing attention: Machine learning models are not static and require ongoing maintenance to ensure they continue to perform well. As data changes over time, models need to be updated or retrained to remain accurate and relevant. There is also a need to ensure that the models are deployed and used appropriately, and that ethical and legal considerations are taken into account.

Future directions for the use of machine learning in environmental aspects include:

- Increased emphasis on interpretability and transparency: As machine learning models become more complex, it will become increasingly important to develop methods for understanding how the models arrive at their predictions. This will be particularly important in fields such as conservation and natural resource management, where decisions based on machine learning models can have significant impacts on ecosystems and biodiversity.

- Integration with other technologies: Machine learning can be combined with other technologies, such as remote sensing and Internet of Things (IoT) devices, to create more powerful and comprehensive environmental monitoring and management systems. For example, machine learning models can be used to analyze satellite imagery and other remote sensing data to detect and monitor environmental changes.
- Focus on real-world applications: There is a need for more research on the practical applications of machine learning in environmental management and decision-making. This includes developing user-friendly interfaces that allow decision-makers to interact with machine learning models and visualize the results.
- Collaboration between industry, academia, and government: The development and deployment of machine learning models in environmental applications require collaboration between industry, academia, and government. This will ensure that the models are scientifically rigorous, scalable, and have real-world impacts.
- Addressing ethical and legal concerns: As the use of machine learning in environmental applications grows, there is a need to address ethical and legal concerns, such as bias, privacy, and data security. This will require the development of frameworks and guidelines for the responsible use of machine learning in environmental management and decision-making.

4.13 Conclusion and Recommendations

In conclusion, machine learning has become an increasingly important tool in addressing environmental challenges. It has proven to be effective in a wide range of applications such as air quality monitoring, water quality prediction, climate change analysis, habitat suitability modeling, and waste management. However, there are still challenges in applying machine learning in environmental aspects such as data quality, transparency, and ethics. To advance the use of machine learning in environmental aspects, there is a need for collaboration between researchers, practitioners, and policymakers. Data sharing and standardization should be encouraged to promote transparency and reproducibility. Ethical considerations and bias in data and algorithms should also be addressed to ensure fairness and prevent unintended consequences. Moreover, there is a need for capacity building and training programs to equip environmental professionals with the necessary skills to use machine learning tools effectively. Public engagement and education should also be prioritized to enhance the understanding and acceptance of machine learning in environmental decision-making. In summary, machine learning has enormous potential to contribute to environmental sustainability and help address environmental challenges. By leveraging the opportunities and addressing the challenges, we can make significant progress in achieving a sustainable future for all.

Recommendations for the improvement and integration of Machine Learning in Environmental Aspect:

Based on the challenges, opportunities, and lessons learned, here are some recommendations for the improvement and integration of Machine Learning in Environmental Aspect:

- Increased collaboration: Collaboration between machine learning experts and environmental domain experts should be encouraged to ensure that the models developed are relevant, accurate, and reliable.
- Data management: Effective data management strategies should be implemented to ensure that data quality is maintained, and data are accessible and available for machine learning applications.
- Transparency and ethics: There is a need for transparency and ethical considerations in the development and application of machine learning models. This includes transparency in the data used, the model algorithms and parameters, and the decision-making process.
- Interdisciplinary education: There is a need for interdisciplinary education and training programs that integrate machine learning and environmental science to equip the next generation of experts with the necessary skills to

address environmental challenges using machine learning techniques.
- Standardization: Standardization of data collection, management, and analysis procedures is essential for the development of accurate and reliable machine learning models.
- Increased funding: Increased funding should be provided to support research and development of machine learning applications in environmental science.
- Continued evaluation: On-going evaluation and validation of machine learning models are necessary to ensure that they remain relevant and accurate in the changing environmental conditions.

By implementing these recommendations, we can improve the effectiveness and efficiency of machine learning applications in environmental science, leading to better environmental management and conservation efforts.

References

1. Kheirandish, S., Taghizadeh-Mehrjardi, R., Salehi, B., Soffianian, A., Land cover and land use change detection based on NDVI using machine learning algorithms. *J. Appl. Remote Sens.*, 13, 2, 024501, 2019.
2. Liu, X., Ma, J., Li, L., Li, Q., Li, X., Li, X., Li, Y., A machine learning-based approach for predicting water quality. *J. Hydrol.*, 566, 643–651, 2018.
3. Chen, J., Li, X., Huang, G., Zhao, W., Liu, X., Urban solid waste generation forecasting using machine learning models. *J. Clean. Prod.*, 212, 1184–1194, 2019.
4. Ager, A.A., Vaillant, N.M., Finney, M.A., A deep learning convolutional neural network can recognize and predict wildfire conditions. *Int. J. Wildland Fire*, 28, 11, 897–903, 2019.
5. Babak, M.V., Fallahpour, M., Borhani, M., Land cover mapping using Random Forest algorithm and remote sensing data. *J. Environ. Manage.*, 233, 457–467, 2019.
6. Deng, Y., Li, Y., Li, H., Li, W., Land use classification based on decision tree algorithm and remote sensing data in Liangshan region, China. *J. Geogr. Sci*, 29, 7, 1127–1140, 2019.
7. Li, C., Li, M., Zhang, H., Huang, L., Cao, W., A comparative study of deep learning and machine learning algorithms for land cover classification from remote sensing images. *Remote Sens.*, 13, 4, 610, 2021.

8. Li, S., Wang, S., Pan, G., Zhang, G., Prediction of soil organic matter using support vector machines and soil spectral reflectance data. *Pedosphere*, 29, 3, 357–369, 2019.
9. Liu, J., Chen, J., Tian, Y., Zhang, M., Pan, X., A machine learning-based approach to predicting water quality in lakes. *Environ. Sci. Pollut. Res.*, 25, 24, 24158, 2018.
10. Wulder, M.A., White, J.C., Gergel, S.E., Hobart, G.W., Hermosilla, T., Mapping and monitoring biodiversity with remote sensing: achievements and challenges. *Remote Sens. Environ.*, 252, 112142, 2021.
11. Ferretti, F., Tabarelli, M., Naves, J., Machine learning-based models to predict habitat suitability for the brown bear (Ursus arctos) in the Italian Alps. *Ecol. Inf.*, 55, 101046, 2020.
12. Norouzzadeh, M.S., Nguyen, A., Kosmala, M., Swanson, A., Palmer, M.S., Packer, C., Clune, J., Automatically identifying, counting, and describing wild animals in camera-trap images with deep learning. *Proc. Natl. Acad. Sci.*, 115, 25, E5716–E5725, 2018.
13. Lin, C.T., Chen, Y.H., Chen, Y.C., Liu, Y.C., Lin, Y.S., Detecting illegal logging activity with a deep learning approach. *Int. J. Appl. Earth Obs. Geoinf.*, 76, 173–181, 2019.
14. Bhandari, R., *et al.*, Machine Learning for Plastic Waste Classification. *Proceedings of the 7th International Conference on Solid Waste Management (ICSW)*, 20192019.
15. Tarek, M., *et al.*, Predictive Modeling of Solid Waste Generation in Urban Areas Using Artificial Neural Networks: A Case Study in Alexandria, Egypt. *J. Environ. Manage.*, 249, 109383, 2019.
16. Anzovino, M., *et al.*, Artificial Intelligence for Plastic Recycling: A Review. *Waste Manage. (Oxford)*, 79, 586–592, 2018.
17. Ding, T., *et al.*, Methane Emission Prediction of a Landfill Using Machine Learning Techniques. *J. Clean. Prod.*, 197, 788–797, 2018.

5

Green Engineering in IoT

Talwinder Singh* **and Davinder Singh**

Department of Mechanical Engineering, Punjabi University, Patiala, Punjab, India

Abstract

Green engineering is the design of processes and products to reduce pollution, promote sustainability, and minimize risks to human health and the environment, without compromising economic efficiency. The Internet of Things (IoT) has a large environmental and economic impact with billions or trillions of interconnected devices communicating over the internet using various sensors. It is well known that each sensor requires a small amount of energy to operate, however with billions of sensors, the energy consumption can be very high. Therefore, it is important to focus on developing efficient IoT technologies and sustainable solutions. This chapter highlights different eco-friendly IoT techniques such as green wireless sensor network (WSN), green machine to machine (M2M), green data center (DC), green cloud computing (CC), and green radio-frequency identification (RFID) to attain green IoT objectives. Green IoT has resulted in the creation of smart technologies that promote environmental sustainability by optimizing use of conventional energy sources, reusing materials, using renewable energy sources and recycling products to establish an eco-friendly and sustainable living space. At the end of chapter, recommendations for raising awareness and potential future directions for green IoT have also been discussed.

***Keywords*:** Green engineering, Internet of Things (IoT), green IoT, green wireless sensor network (WSN), green machine to machine (M2M), green data center (DC), green cloud computing (CC), green radio-frequency identification (RFID)

**Corresponding author*: tp_tiet@yahoo.co.in

Chandan Deep Singh, Talwinder Singh and Davinder Singh (eds.) Digitization and Manufacturing Performance: An Environmental Perspective, (129–148) © 2025 Scrivener Publishing LLC

5.1 Introduction

Green engineering is the design of processes and products to reduce pollution, promote sustainability, and minimize risks to human health and the environment, without compromising economic efficiency. Green engineering encompasses the idea that decisions to protect human health and the surroundings; can have the greatest impact and benefit when implemented in the early stages of design and development of a product or process [1, 2]. The Internet of Things (IoT) describes a network of physical devices (things) integrated with software, sensors, and other technologies to connect and exchange information with other devices and systems on the internet. These tools range from home appliances to advanced business tools [3]. The Internet of Things (IoT) is an innovative technology that is driving the advancement of telecommunications and improving the quality of life of people around the world. IoT also has the potential to significantly increase global trade. It is expected that the global economic impact of the Internet of Things will be between USD 2.7 trillion and 6.2 trillion by 2025, because IoT devices are a critical component of emerging applications and playing a central role in the widespread adoption of machine-type communication [4, 5]. IoT technology utilizes built-in sensors, communication hardware and software to gather and transmit data produced by its usage. IoT devices provide this information through IoT gateways or platforms or other IoT devices [6].

5.2 IoT Data Types

IoT data types are classified as:

- Status data: Status data (also called as raw data) tells the status of a system or a device.
- Automation data: Data produced by automated systems and devices such as automated lighting system and smart thermostats is known as automation data.
- Location data: This type of data communicates geographical position of the system or device mainly applied in warehousing, manufacturing, and logistics monitoring [6].

5.2.1 IoT Data Value

IoT data has been used in many ways to attain different organizational objectives as discussed below:

1. Increase productivity and efficiency
With data from connected systems, you can access bottlenecks and opportunities to improve efficiency. You can also run tests to see which methods provide the most useful results.

2. Understand about equipment and systems
Find out how your equipment and system are performing and which machines need maintenance or are having problems. The important thing is to prevent many issues from happening and to address them quickly when they arise.

3. Create new business models
IoT data assists you to identify how customers use your products. Using this information, you can build new business models and generate revenue.

4. Enhancing customer experience
By improving performance, you will improve the customer experience. You can also uncover opportunities for improvement using data from your product.

5. Better marketing
When you understand how customers use your products and how different segments use them in different situations, you can target your marketing communications correctly.

6. Enhance security
With IoT security tools, you can augment the security of buildings, devices, and other assets. You can also record where employees are at any given time, helping you recognizes internal security violations as soon as they occur [6].

Figure 5.1 shows IoT application areas such as educational institutions, warehouses, agricultural activities, offices, cities, factories, our homes, transports, industries, hospitals and healthcare facilities and services [7].

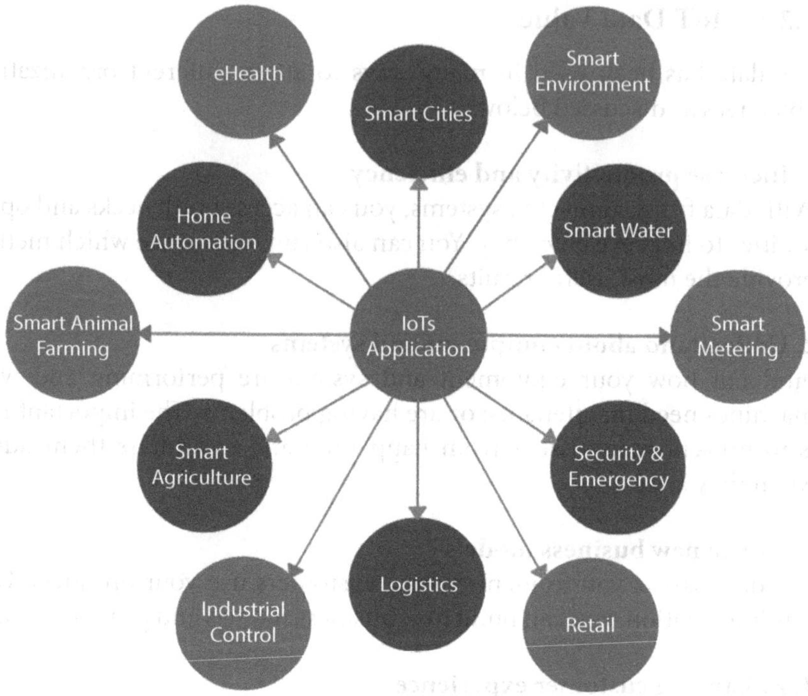

Figure 5.1 IoT applications [7].

5.3 What is Green IoT?

Green Internet of Things (Green IoT) refers to energy efficient practices (hardware or software) applied by the Internet of Things to help reduce the greenhouse effect of existing applications and services, or to reduce the environment impacts of the IoT facility itself. Initially, the use of the Internet of Things helps to reduce the greenhouse effect, while later; the IoT greenhouse footprint will improve further. The entire lifecycle of the green IoT should revolve around green utilization, green production, green design, and eventual green disposal/recycling without impacting the environment [8, 9]. However, in reality green IoT implementation is only possible by the cooperation of several enablers as listed in Figure 5.2. Green IoT is needed in six main areas, identified based on their effect on the surroundings as illustrated in Figure 5.3.

Different communications systems, computing devices, and network architectures have adopted green IoT design principles, including ecology, energy use and carbon emissions as critical parameters. The following

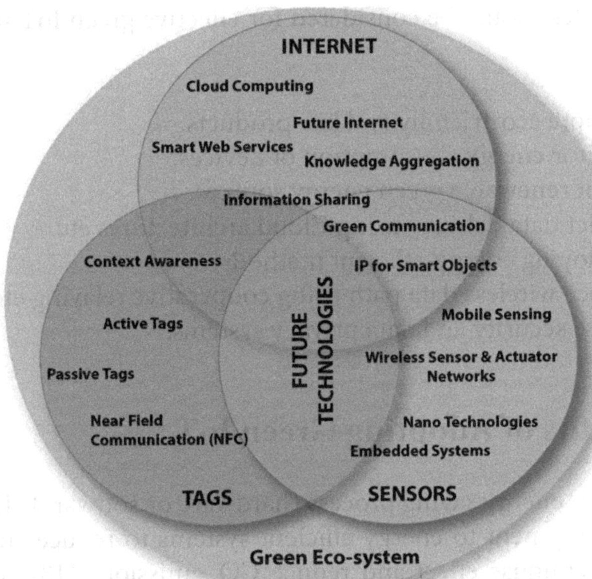

Figure 5.2 Green IoT enablers [8, 10].

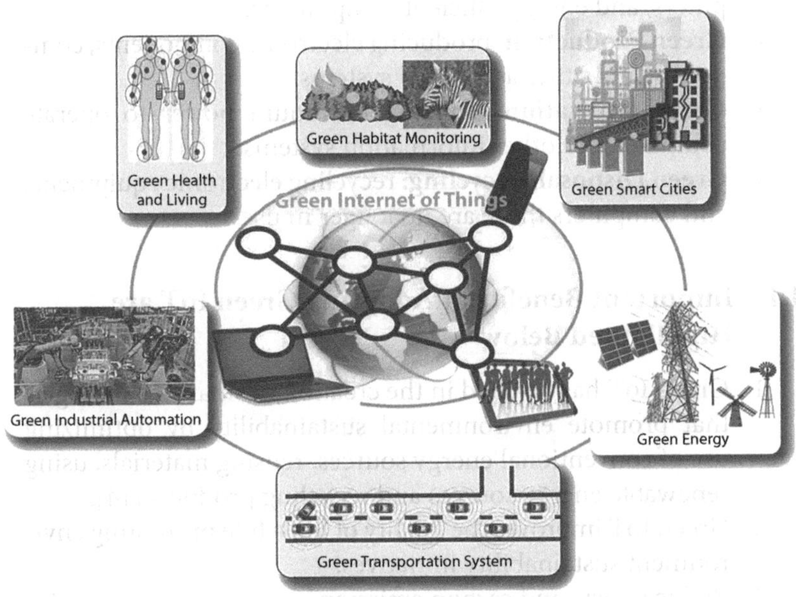

Figure 5.3 Green IoT applications [8, 11].

design principles should be considered for effective green IoT implementation [7, 12]:

- Promote eco-friendly and bio-products.
- Monitor energy consumption of devices.
- Use of renewable green energy sources.
- Predict data delivery using Cloud architectures etc.
- Employing energy-efficient methods.
- Reduce wireless data path using cooperative relaying etc.
- Utilize security and data privacy systems.

5.4 Benefits of Adopting Green IoT

Green IoT is an energy saving process (hardware or software). This means connecting equipment to energy efficient systems to reduce energy consumption, greenhouse effect and reduce CO_2 emissions [13]. To achieve green IoT product, it must go through green design, green production, green utilization, and green disposal/recycling systems.

- **Green Design:** designing servers, cooling equipment, computers, and energy-efficient components.
- **Green Production:** producing electronic components, computers, and other associated systems.
- **Green Utilization:** utilizing minimum power to operate computer and other information systems.
- **Green Disposal/Recycling:** recycling electronic equipments and computers those are no longer in use.

5.4.1 Important Benefits of Adopting Green IoT are Highlighted Below

1. Green IoT has resulted in the creation of smart technologies that promote environmental sustainability by optimizing use of conventional energy sources, reusing materials, using renewable energy sources and recycling products [14].
2. Green IoT improves the quality of life while promoting environment sustainability initiatives.
3. Reduce waste and carbon emission.

4. Design smaller nano-scale energy efficient devices by employing nanotechnology.
5. Reduce costs.
6. Gain business.
7. Enhanced operational efficiency.
8. Results smarter, eco-friendly, and sustainable living space.
9. Improves company's image.

5.5 Green IoT Components

Figure 5.4 highlights the five major Green IoT components to achieve sustainable environment objectives.

5.5.1 Green Wireless Sensor Network (WSN)

A WSN is a group of sensor nodes that are wirelessly interconnected with at least one central node, called a base station (BS). The base station not only manages the network centrally, but also communicates with end users and/or other networks [16]. WSN has many aspects of energy management strategies: conservation, harvesting and transfer.

Energy conservation technology aims to extend the life of the network by reducing power consumption while WSN continues to work as needed [17]. Energy saving often means to minimize communication costs of the nodes [18]. The energy harvesting strategy is to increase the energy

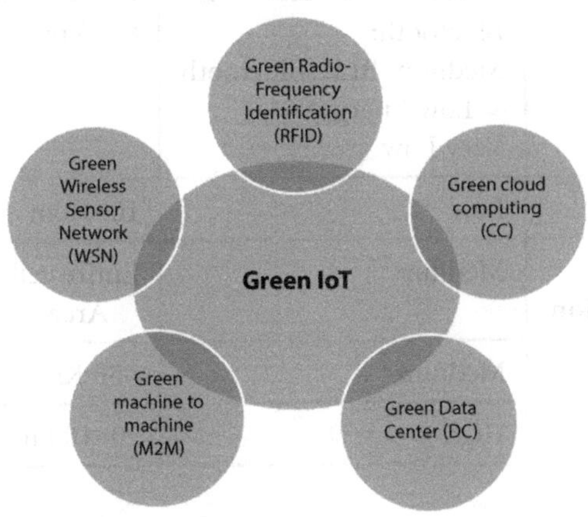

Figure 5.4 Green IoT components [15].

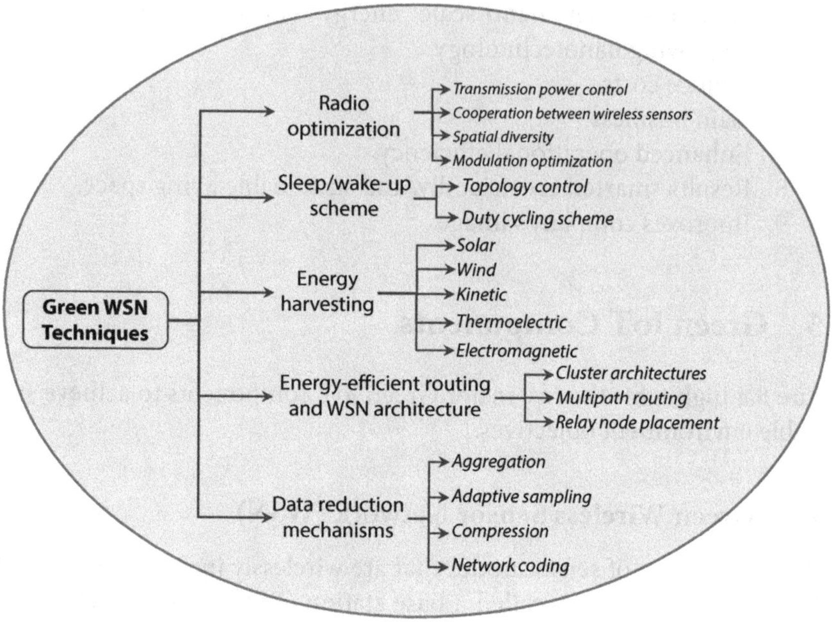

Figure 5.5 Energy-efficient WSN techniques [4].

Table 5.1 Energy consumption for different wireless technologies for IoT applications [4].

Different wireless technologies	Energy consumption	Transmission range	Cost
LoRa	Very Low	<30 Km	High
Bluetooth	Bluetooth: Medium; BLE (Bluetooth Low Energy): Very Low	8–10 m	Low
LR-WPAN	Low	10–20 m	Low
Mobile Communication	Medium	Entire cellular Area	Medium
WiMAX	Medium	<50 Km	High
WiFi	High	20–100 m	High

for the nodes. The solar energy, wind, thermal energy etc. can be obtained from the external environment. This technology converts energy from the environment into electricity at the node. A major limitation of this system is that energy sources are not constantly available and thus it is important to preserve the harvested energy using capacitors or rechargeable batteries.

Energy transfer is another emerging tool for extending the life of the networks. In this system, energy rich nodes transfer energy wirelessly to the energy deficient nodes in the same network. Figure 5.5 summarizes different energy efficient WSN techniques and Table 5.1 shows energy consumption for different wireless technologies for IoT applications.

5.5.2 Green Machine to Machine (M2M)

Machine-to-machine, or M2M, is a general term that can be used to describe any system or technology that allows connected devices to exchange data and perform actions without human assistance. The potential advantages of M2M are given below [19]:

- Reduce costs by optimizing device maintenance and downtime.
- Enhance revenue by providing new business opportunities for servicing products in the field.
- Improve customer service by proactively servicing and monitoring equipment before it stops working or only when it is required.

Automation and efficiency are the two main goals of M2M technology. It is used to facilitate communication, to provide remote monitoring solutions, and finally to ensure smooth operation of the interface. This leads to greater productivity, easier decision-making processes, and a better return on investment [20].

Any M2M communication system depends on the host components within a flexible framework. Some of the most relevant components include:

- Mobile and fixed network services (such as wireless routers, and Bluetooth-compatible devices).
- Hardware components (i.e. microprocessors, wireless sensors, data endpoints, and data integration point servers).
- M2M applications (such as asset management software solutions).

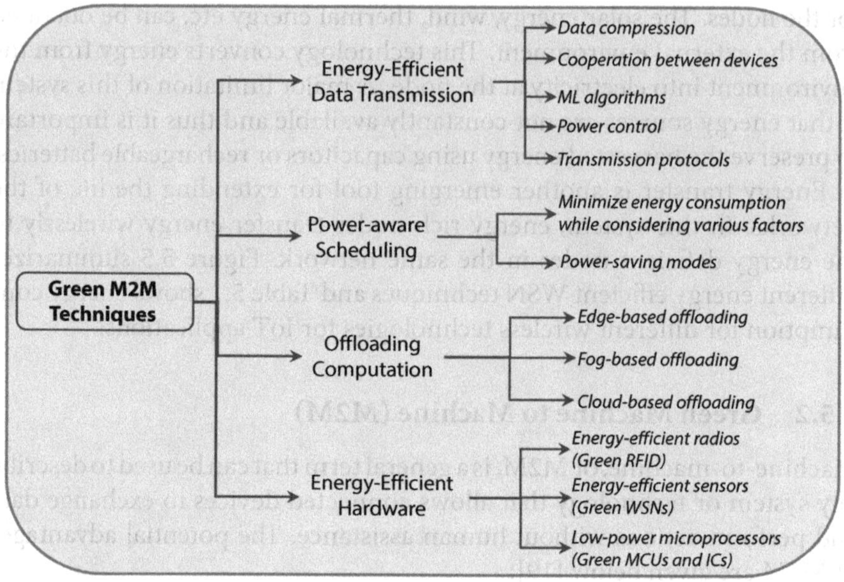

Figure 5.6 Energy-efficient M2M techniques [4].

- System integration and consulting services (such as digital pipeline management software, computing wireless networks to provide higher levels of quality assurance).

Figure 5.6 depicts important energy efficient M2M techniques. However, the energy saving for each technique depends on certain key factors such as type of M2M communication system, operating environment and the device software and hardware configurations.

5.5.3 Green Data Center (DC)

Green data center defined as a place that hosts servers to manage, store, and circulate data that has been considered to be highly energy efficient and reducing its impact on environment [21]. The development of green data centers has become necessary to attain green IoT objectives as data centers are the key structure for global sustainable development. Green data centers employs following strategies and technologies [22]:

- Low-emission building materials, carpets and paints.
- Minimized building footprints.
- Sustainable landscaping.

- Catalytic converters on backup generators.
- E-waste recycling.
- Alternative energy, such as photovoltaic technology and evaporative cooling technology.

Adopting green data centers provide the following benefits [22]:

- Decrease in carbon emissions.
- Reduces operating costs.
- Reduces space requirements.
- Decreases waste output.
- Emphasizes use of renewable and sustainable data center.
- Reduces electricity consumption.

5.5.3.1 *Ways to Achieve a Greener Data Center [23]*

Cloud computing, which became popular in the mid-2000s, revolutionized the data center industry. Cloud computing promises to be more secure than other methods of data storage and processing, while also provides mobility, security, cost-savings, and scalability benefits. Cloud operators can efficiently reduce energy consumption and other data that often increases the carbon footprint of data centers.

The proliferation of cloud services has increased the need for large data centers around the world. In response to the growing concern of governments, industry regulators, and consumers about the security of data centers and cloud operators, many organizations have set a goal of being carbon neutral by 2030.

The entire data center industry uses around 150 billion kWh of electricity each year, which is nearly 3% of global energy consumption. Therefore, data centers should work as efficiently and sustainably as possible to eliminate environmental damage. Following are the ways to achieve greener data center:

1. Upgrade to New Equipment

Although the equipment is not considered to be faulty as its operational efficiency founds to be optimal, but older the equipment, the less energy efficient it may be. This problem is most serious in older data centers where important electrical and electronic equipment such as uninterruptible generators, transformers appear obsolete and transformers are nearing the end of their recommended service life. Although regular maintenance and repair can improve equipment performance, over time equipment becomes faulty and more expensive to maintain. It is important to replace not only larger, important equipment, but also small, forgotten items such as capacitors and circuit breakers.

2. Reduce Bypass Airflow

Another part of the data center that contributes to energy efficiency is air conditioning. Data center cooling is estimated to account for up to half of all energy consumption in data centers, so optimizing airflow is important for energy efficiency. Bypass airflow is air returning to the cooler without removing heat. This leads to a reduction in cooling capacity, increased cooling costs and increased temperatures. Therefore, bypass airflow reduces the efficiency of the cooling system and requires additional cooling to remove heat from the IT equipment. Bypass airflow can be controlled via louvers to improve airflow, reduce server inlet temperature, and increase CRAC (computer room air conditioning) return air temperature. This solution can reduce waste energy, thus reducing energy costs by nearly 1–2%.

3. Shift to Economy Mode (ECO)

Although an uninterruptible power supply (UPS) can operate in different modes to reduce energy waste, but it also waste a lot of energy. Economy (ECO) mode is an option designed to reduce losses associated with inefficient energy usage, and it involves incoming power passing directly through the UPS. In this mode, the server runs on mains power but is protected in the event of a power outage.

4. Employ Server Virtualization

Server virtualization involves using physical servers as a pool of computing capacity. Servers are divided into several "virtual machines" that operate like separate physical machines, as they facilitate the operation of many systems and applications. This is one of the best ways to increase server utilization, consolidate space and equipment, and reduce energy consumption as it reduces the number of physical servers needed in the data center. Because fewer servers are used, power and cooling are not needed, thus reducing energy consumption.

5. Reuse or Recycle Equipment

In addition to traditional paper and plastic recycling, warehouses must be able to recycle and reuse their equipment to reduce unnecessary emissions left in place during production. Recycling is also a great way to reduce your data center's carbon footprint. Figure 5.7 illustrates different actions needed to reduce e-waste.

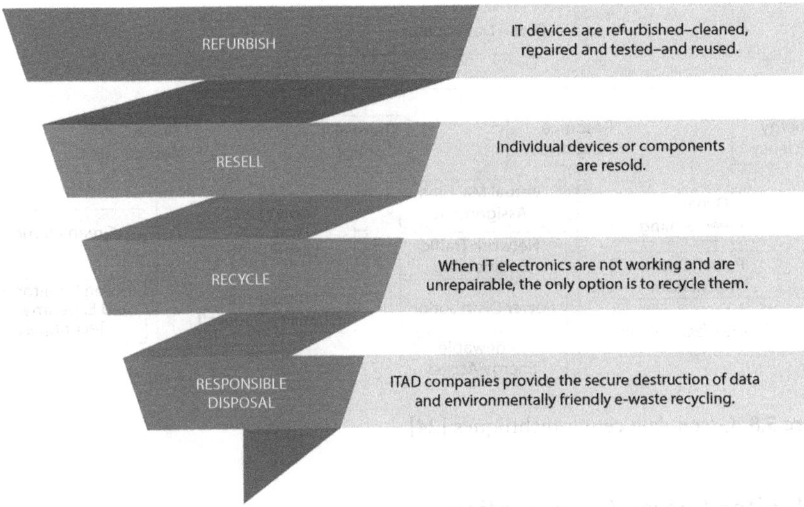

Figure 5.7 Actions to reduce e-waste [22].

6. Implement Green Certification Systems

Many green building certification programs are available to help teach, present and document efforts towards more efficient, effective storage facilities. The most important is the Leadership in Energy and Environmental Design (LEED) certification, established by the US Green Building Council (USGBC). It is a rating system with certificates such as Silver, Gold and Platinum for the design, construction, operation and maintenance of green buildings. Platinum certified centers are data centers with the highest level of environmental protection and efficient use of resources. Some of the features of LEED certification include cooling, reduced energy consumption, energy efficiency, renewable energy and green buildings. It is also an opportunity to gain a competitive advantage over other data centers, as customers who value environmental sustainability will prefer LEED-certified data center over a competitor.

7. Use Renewable Energy

Wind turbines, solar panels, and hydroelectric power plants are examples of renewable energy sources that can replace fossil fuels. Next-generation renewable energy could be even cheaper than the cheapest fossil fuel generator option. Aside from the cost of renewable energy, one of the most obvious benefits is that sources are very unlikely to exhaust or become unavailable.

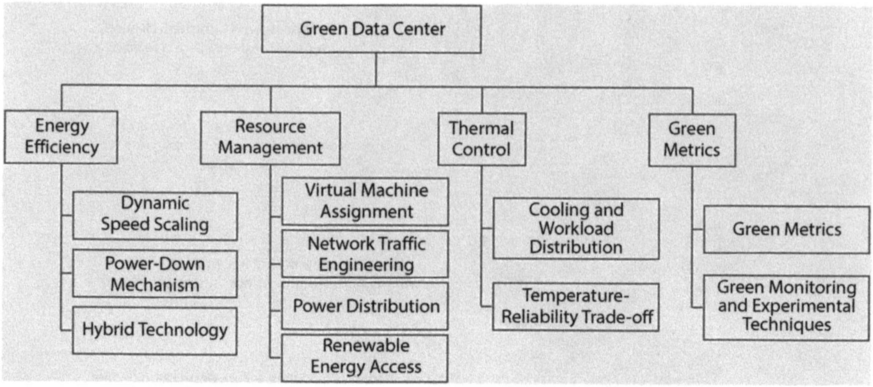

Figure 5.8 Green data center techniques [24].

8. Monitor Energy Consumption

Power monitoring is an essential part of keeping your data center as green as possible. You can also set a timer for the data to be recorded, so you will be notified when the timer is over. Figure 5.8 highlights different green data center techniques.

5.5.4 Green Cloud Computing (CC)

Green cloud computing is a cloud infrastructure where data centers can use resources more efficiently and have a lower carbon footprint. In the EU, the European Commission is taking broad steps to force data centers to adopt eco-friendly practices, aiming to be carbon neutral by 2030 [25].

Cloud computing can be implemented through:

- In-house datacenters.
- Third-party services owned and operated by Amazon Web Services (AWS), Microsoft Azure, and Google Cloud Platform.
- Hybrid clouds (combination of third-party and In-house datacenters).

While all three types of cloud computing theoretically achieve green cloud, greater energy efficiency and reduced carbon footprint can be attained in cloud computing. This is because large players like Microsoft can take advantage of more market opportunities to make better profits and buy more renewable energy in their IT operations, IT products and data centers.

5.5.4.1 Green Cloud Computing Objectives

Green cloud computing is implemented to accomplish the following objectives:

(1) Reduce Hazardous Materials
Network equipment, physical servers, and other equipment used in data centers may contain hazardous materials such as lithium, mercury, lead, and PVC plastic, which can be hazardous to the environment if disposed of as e-waste. By reducing the number of hardware devices in your data center, you can reduce the amount of e-waste released to the environment after these devices reach the end of their life.

(2) Improve Energy Efficiency
Data center hardware consumes energy throughout its lifetime. The more energy they use, the more carbon dioxide and other carbon monoxide are released into the environment (i.e. carbon footprint). You can reduce your carbon footprint by choosing energy-efficient equipment and technology.

(3) Encourage Product Recycling and Reuse
In 2019, the world generated nearly 53.6 metric tons of e-waste. This number will only increase as people and organizations rely more on technology. So, every time you delay the disposal of products, you also delay the disposal of e-waste. One way is to use chemicals that allow you to reuse or recycle existing equipment.

5.5.4.2 Green Cloud Computing Benefits

When you run your business in the cloud instead of using on-premises data centers, there are several things that can benefit the environment, as described here:

(a) Decreased Energy Consumption
In-house data centers use a lot of energy. This energy goes to physical servers, storage devices, network equipment, air conditioners, etc. used to provide power. Cloud environments also have these components but they are more energy-efficient. This is owing to utilization of:

- Server consolidation via virtualization - Reduces number of physical servers.

- Automation and dynamic provisioning - This prevents over provisioning of virtual machines, a common problem in traditional data centers that causes wasted storage space and energy.
- Multitenancy - Allows multiple organizations to share resources without disrupting operations.

(b) Reduced Greenhouse Gas (GHG) Emissions

The equipment that makes up the data center (e.g., network devices, physical servers, air conditioners) generates greenhouse gases throughout their lifetime – from manufacturing, assembly, transportation, operation and also during their disposal. As the similar components in cloud environments are more efficient, the total greenhouse gas emissions are greatly reduced.

(c) Dematerialization

Dematerialization is the practice of replacing physical objects with their virtual equivalents. For example, replace the physical server with the virtual server or replace the DVD player with a cloud-based video streaming service. This not only reduces the overall carbon footprint of the organization, but also reduces e-waste.

(d) Utilize Renewable Energy Sources

Large cloud providers like Google Cloud implements more renewable energy to be purchased directly from the energy certificate or from a public utility that generates electricity directly from renewable sources such as wind and solar. Moreover, many aim to generate 100% of their electricity from renewable sources in the future.

5.5.5 Green Radio-Frequency Identification (RFID)

Green Radio Frequency Identification (RFID), one of the Green IoT components, refers to a wireless system with two components: a tag and a reader. A reader is a device with one or more antennas that emit radio waves and receive signals back from RFID tags. Tags, passive or active, use radio waves to transmit themselves and other information to nearby readers. Passive RFID tags are powered by a reader and do not require batteries, whereas active RFID tags are operated by batteries [26, 27]. RFID tags can store a variety of information, from barcodes to multiple pages of information. Readers can be mobile, to be carried by hand or mounted on a pole or overhead. Reader systems can also be placed in cabinets, rooms or buildings. Figure 5.9 illustrates different applications of RFID tag.

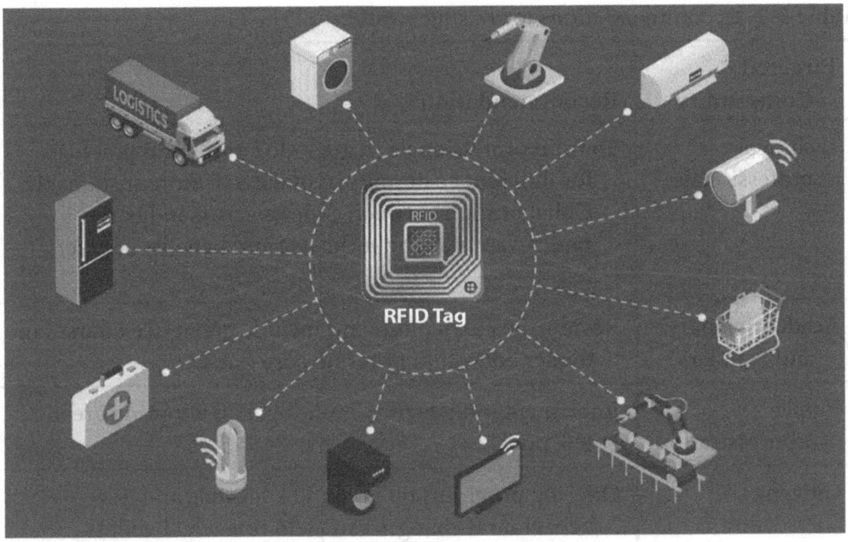

Figure 5.9 Applications of RFID tag [28].

Green RFID can be a powerful tool for organizations to accomplish sustainable development such as [29]:

- Green RFID can reduce the need for papers and receipts, and thus helps in conserve resources and reduce waste.
- Green RFID monitors the humidity and temperature of perishable products to prevent spoilage and waste.
- It is also possible to monitor the origin and quality of raw materials and products, ensuring integrity and fair trade.
- Can assist circular economy models by providing information on the life cycle and condition of products such as reuse, repair and recycling.
- Green RFID can improve energy use and carbon footprint by optimizing transportation, logistics and products.

5.6 Recommendations for Raising Awareness and Future Research Directions

Table 5.2 shows suggested action plans for creating awareness of green IoT among the society. Correct planning and joint activities in energy awareness programs are effective in increasing user awareness and promoting

Table 5.2 Recommendations for raising awareness [7].

Targeted Community	Recommendations
Technical professionals	To train professionals in green IoT skills and practices for different customers to make IoT more sustainable. Collaboration between experts across industries to raise awareness about the use and benefits of green IoT.
Academicians and Scholars	Review the curriculum and include green IoT courses in higher education programs or courses.
Public administration	Implementation, facility building, and adopting green IoT policies.
Citizens	Offering programs on green IoT lifestyle and how behavioural changes can yield financial benefits.
Future generation	The importance of teaching young people how to use green lifestyles like sustainable IoT. Importance of circular economy and how to reduce energy consumption. By teaching and raising awareness to the younger generation, will create higher value in the future and they will see it as lifelong learning.

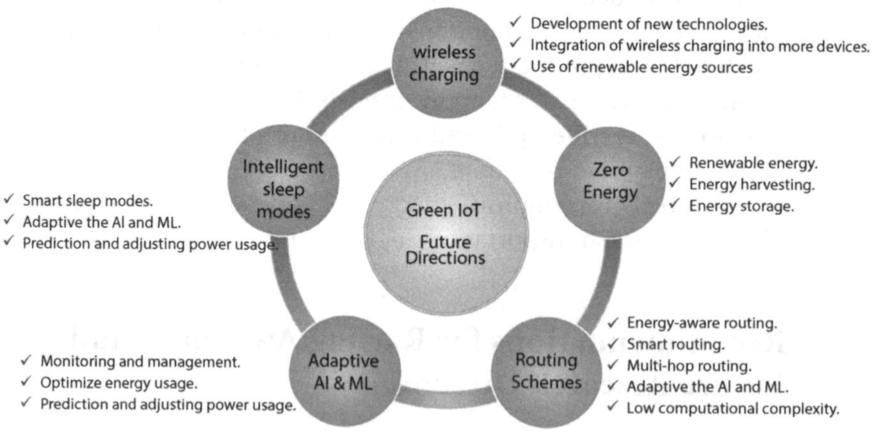

Figure 5.10 Potential future directions for green IoT [4].

energy saving practices. Create energy awareness in schools by using promotional tools such as videos, posters, educational games to raise student awareness. Workplace environmental awareness includes teaching employees energy-saving practices, encouraging employees to know about energy-saving places, posting energy-related posters in office buildings, etc. Figure 5.10 highlights green IoT future directions for eco-friendly and sustainable environment.

References

1. See https://www.epa.gov/green-engineering/about-green-engineering.
2. Oncel, S.S., Green Engineering and the Realization of the Sustainable Future Perspective, in: *A Sustainable Green Future*, pp. 3–21, Springer Cham, Switzerland, 2023.
3. See https://www.oracle.com/in/internet-of-things/what-is-iot/#:~:text=What%20is%20IoT%3F,and%20systems%20over%20the%20internet.
4. Alsharif, M.H., Jahid, A., Kelechi, A.H., Kannadasan, R., Green IoT: A Review and Future Research Directions. *Symmetry*, 15, 3, 757, 2023.
5. Al-Fuqaha, A., Guizani, M., Mohammadi, M., Aledhari, M., Ayyash, M., Internet of things: A survey on enabling technologies, protocols, and applications. *IEEE Commun. Surv. Tutor.*, 17, 4, 2347–2376, 2015.
6. See https://www.snowflake.com/guides/what-iot.
7. Albreem, M.A., Sheikh, A.M., Alsharif, M.H., Jusoh, M., Mohd Yasin, M.N., Green Internet of Things (GIoT): Applications, Practices, Awareness, and Challenges. *IEEE Access*, 9, 38833–38858, 2021.
8. Shaikh, F.K., Zeadally, S., Exposito, E., Enabling Technologies for Green Internet of Things. *IEEE Syst. J.*, 11, 2, 983–994, 2017.
9. Murugesan, S., and Gangadharan, G.R., Green IT: An Overview. Harnessing Green IT: Principles and Practices, *IEEE*, 1–21, 2012.
10. Atzori, L., Iera, A., Morabito, G., The Internet of Things: A survey. *Comput. Netw.*, 54, 15, 2787–2805, 2010.
11. Vermesan, O., EU IoT Strategic Research Agenda 2012, in: *The Internet of Things 2012—New Horizons*, Halifax, European Research Cluster, U.K, 2012.
12. Maksimovic, M., Greening the future: Green Internet of Things (G-IoT) as a key technological enabler of sustainable development. *Internet Things Big Data Anal. Toward Next-Generation Intell.*, 30, 283–313, 2018.
13. See https://www.speranzainc.com/green-iot-sustainable-design-and-technologies/#:~:text=Green%20IoT%20is%20an%20energy,minimizing%20the%20emission%20of%20CO2.
14. See https://www.infoq.com/articles/green-iot-energy-sustainability/.

15. Ahmad, R., Asim, M., Khan, S.Z., Singh, B., Green IoT — Issues and Challenges. *Proceedings of 2nd International Conference on Advanced Computing and Software Engineering (ICACSE)*, 378–382, 2019.
16. López-Ardao, J.C., Rodríguez-Rubio, R.F., Suárez-González, A., Rodríguez-Pérez, M., Sousa-Vieira, M.E., Current Trends on Green Wireless Sensor Networks. *Sensors*, 21, 13, 4281, 2021.
17. Anastasi, G., Conti, M., Di Francesco, M., Passarella, A., Energy conservation in wireless sensor networks: A survey. *Ad Hoc Netw.*, 7, 3, 537–568, 2009.
18. Raghunathan, V., Schurgers, C., Park, S., Srivastava, M.B., Energy-aware wireless microsensor networks. *IEEE Signal Process Mag.*, 19, 2, 40–50, 2002.
19. See https://www.techtarget.com/iotagenda/definition/machine-to-machine-M2M#:~:text=Machine%2Dto%2Dmachine%2C%20or,the%20manual%20assistance%20of%20humans.
20. See https://toolsense.io/glossary/m2m/#:~:text=M2M%20technology%20involves%20the%20automatic, supply%20chain%20management%20(SCM).
21. See https://www.araner.com/blog/what-is-a-green-data-center.
22. See https://www.techtarget.com/searchdatacenter/definition/green-data-center.
23. See https://www.sunbirddcim.com/blog/8-ways-ensure-greener-data-center.
24. Jin, X., Zhang, F., Vasilakos, A.V., Liu, Z., Green Data Centers: A Survey, Perspectives, and Future Directions. *Comput. Sci. Dist. Parallel Cluster Comput.*, 1–20, 2016.
25. See https://www.parallels.com/blogs/ras/benefits-green-cloud-computing/.
26. Yusoff, Z.Y.M., Ishak, M.K., Alezabi, K.A., , The Role of RFID in Green IoT: A Survey on Technologies, Challenges and. *Adv. Sci. Technol. Eng. Syst.*, 6, 1, 17–35, 2021.
27. See https://www.fda.gov/radiation-emitting-products/electromagnetic-compatibility-emc/radio-frequency-identification-rfid#:~:text=Radio%20Frequency%20Identification%20(RFID)%20refers,back%20from%20the%20RFID%20tag.
28. See https://www.softwebsolutions.com/resources/IoT-adoption-with-RFID-technology.html.
29. See https://www.linkedin.com/advice/3/how-can-rfid-support-sustainability#:~:text=RFID%20can%20be%20a%20powerful,preventing%20spoilage%20and%20food%20waste.

6

Green Engineering in Product Development

Harpreet Sharma[1]*, Chandan Deep Singh[2] and Kanwaljeet Singh[3]

[1]*Yadavindra Department of Engineering, Guru Kashi Campus, Talwandi Sabo, Punjab, India*
[2]*Punjabi University, Department of Mechanical Engineering, Patiala, India*
[3]*Punjab State Aeronautical Engineering Colleges, Patiala Civil Aerodrome, Patiala, India*

Abstract
Green engineering, also known as sustainable engineering or environmentally conscious engineering, is an approach to product development that aims to minimize the environmental impact throughout the entire product lifecycle. This abstract provides an overview of the principles and practices of green engineering in product development. It highlights the importance of considering environmental factors, such as resource efficiency, pollution prevention, and waste reduction, from the initial design phase to the end-of-life disposal. The abstract also discusses the benefits of adopting green engineering practices, including cost savings, regulatory compliance, and enhanced brand reputation.

***Keywords*:** Eco-design, sustainable materials, lifecycle assessment (LCA), circular economy, cradle-to-cradle (C2C), biomimicry, renewable resources, waste reduction

6.1 Introduction and Meaning

Green engineering deals with the design ideas, commercialization of the products, use of the available processes and available resources in a way that reduce environmental threats, enhance the product sustainability, and to

**Corresponding author*: harpreetsharma3211@gmail.com

Chandan Deep Singh, Talwinder Singh and Davinder Singh (eds.) Digitization and Manufacturing Performance: An Environmental Perspective, (149–166) © 2025 Scrivener Publishing LLC

keep an eye on the pollution caused and treats to human health without compromising on the economics feasibility and overall efficiency of the product. The term Green engineering also known as "Green Technology," or "Clean Technology" [1]. Green engineering tries to adapt the innovative designing process for the products to achieve the following fundamental goals.

- To decrease in the amount of pollution caused by the processing plants in the form of industries, as the demand for the products increase, e.g., to no of vehicles are increasing at very rapid rate every next year.
- To minimize the operator's exposure to certain hazards i.e., to keep an eye on the workplace safety.
- To effectively use the available resources in the form of raw materials and energy throughout the lifecycle and design process of the desired product. Most of the energy available is utilized by the industrial sector among transportation, commercial, and residential sectors [2].
- The product should be feasible in terms of the finance i.e., to keep an eye on the economic feasibility of the product. Generally, the cost of the product increases with quantity as well as quality.
- The designed or manufactured product should be practicable in the respective function or objective for which it is designed or manufactured.
- To minimize the wastes, enhance the recycling, and approaching the sustainability.

The field of Green engineering has turned out the attention of design and manufacturing engineers towards itself due to its much serious positives obtained because of implementing the fundamental principles and standards of Green technology and engineering [3]. The new inventions and advancements in almost every sector have also put some pressure on the planet, Earth. The field of automobile industry has put pressure on the planet in terms of the harmful and poisonous gases emitted by the automobiles as well as the manufacturing plants of these automobiles. In most of the cases this pressure on our planet comes out to be in the form of global warming and greenhouse gases. Serious attention needs to be given to this problem and to introduce some innovative techniques to affectively use the natural resources and the energy without causing threats to our planet and to the kind humanity living on the face of this planet.

Following in the chapter various innovative and incentive design and manufacturing ideas have been discussed to produce the qualitative products with the help of processes using throughout their life cycle which are effective in terms of performance as well as environmentally friendly.

6.2 Principles of Green Engineering

The principles of green engineering provide a strong background for the engineers to involve in the design and manufacturing of new materials, products, operating processes, and complex systems that are beneficial to human health and the environment [4]. The measurement of these principles in products varies from product to product depending on the desired need. The fundamental principles of green engineering are given below.

1. Inherent rather than circumstantial:
The effects of the hazardous substances which tend to cause harm either by toxicity, physical damage, or global can be minimized through investing time, capital, material, and extra energy resources but this in again not economical as well as not environment friendly. Rather the design and manufacturing engineers should analyze the fundamental nature of the selected raw materials and energy resource as an input as the prior step and to ensure that they are kind and neutral to the environment. The designers also need to analyze the process on molecular level to make the product sustainability easy.

There are cases in which the inputs selection has certain associated hazards, these hazards can be removed from the process path during the process at the stage of purification or separation, but this process also requires constant assessment and monitoring. Also, the disposal of the hazards to some permanent off-site place is also required. The other alternative to cope with this problem is to incorporate the associated hazard into the final product and to recycle the hazard again and again, but this incorporated hazard in the final product has the risks of release through some accidents, spills, and leaks [3].

2. Prevention instead of treatment:
The concept of "zero waste" is almost contrary to the fundamental principles of thermodynamics and entropy generation. The laws of thermodynamics states that no single process is 100% efficient and that there is necessary destruction of some part of the available energy in different forms. Also, there is no waste associated with the input material or operating process but is associated with

the misuse of these ingredients. A process is said to have associated waste when it does not deliver the benefits for which it has been installed. Regardless of the nature of the waste produced, the handling and disposal of the waste takes some human effort, consumes time, and energy. Also, it needs additional costs for the monitoring and control. The waste factor is thoughtlessly incorporated into the input design and this is according to the basic concept that, "inputs are designed to be part of the output." This is in accordance with the "atomic economy," which can be extended to "material economy" on large scale. The solution to this problem is to choose the idealized processes for the design and manufacturing, e.g., the use of fusion and fission reactions for the power generation as compared to the use of fuels which ultimately give rise to global warming and green house effects. The scope of nuclear energy needs to be revolutionized because of the harms caused by the fossil fuels and much of the portion of today's world is dependent on the energy consumption based [3].

3. Design for separation and purification:
The input material preparation process is usually associated with certain impacts which are not kind to the environment. The separation and purification process of the starting materials usually require hazardous chemical substances which are hazardous in nature, or some other alternative techniques are required which consumes lot of the energy in the forms of heat and pressure. This can be avoided by the designs of products containing the essential chemical properties of volatility and solubility to carry out the separation and purification of the input material in a very natural way, so that no negative impacts on the environment in the form of wastes disposal, dangerous gases emissions, etc. will be created. The cluster of unwanted parameters for the disposal of, say wastes which are time consumption, money requirements, resources for the disposal, transportation, and monitoring can be avoided by incorporating the desired properties in the components from which products, processes, and systems made. The use of reversible fasteners, or non-permanent joints in the design considerations provide a way for the recycling and reuse of the materials. The design should give keen concern to the up-front design to minimize the material and energy usage, e.g., the purification and separation process of column chromatography and purification require large amount of the hazardous substances and energy, respectively [5].

4. Optimization of mass, space, energy, and time:
Many of the products, operating processes, and integrated systems require more power, materials, time, and space that the ultimate requirements resulting in inefficiency of the overall design. The wastage of the resources can also

be observed when the product is not operated at its full efficiency capability. The design tools of optimizing the time and space should also be used in integration with the optimizing techniques for the materials and energy to enhance the product sustainability. The use of microreactors techniques in processing industries is innovated design idea for materials, energy, space, and time optimization. The traditional techniques of reactions in processing industries in the form of vessel reactors or batch reactors require very large amount of the input that that used the reactions. These innovative techniques of micro reactors in the processing industries tend to enhance the production quantity, and efficiency of the process using very low volume of the material at the input of the process. The design and manufacturing engineers should try to adopt these optimization techniques, e.g., replacing batch or vessels reactors by spinning disk reactors, replacing paints by powder coatings, printed media by digital information's, and eco-industrial plants to diminish poorly planned and auto dependent development .

5. Output pulled processes over the input pushed processes:
This principle of sustainable design and manufacturing is analogous to Lechatelier's principle in chemical composition or chemical synthesis, i.e., applying stress to the process (in the form of providing energy or material) will try to create balance between the inputs and outputs. The continuous production in output will tend to maximize the materials and energy intake at the input of the manufacturing process, also leading to the maximization of space and time constrains. This maximization of the resources at the input of process leads to costly, inefficient, and unsustainable products production. The designers should try the minimize the quantity at output side, and this will pull the limited resources available at the input towards the production machinery. To put an end to this problem of maximizing the resources at the input, the design and manufacturing engineers and analysts should try to adopt the Just in Time (JIT) manufacturing process rather than continuous production process, also known as "Push vs Pull" manufacturing process. It means that the end user of the product is also the final purchaser of the product, and limited amount of the manufacturing is to be performed for the specific need. This will ultimately save the resources, operating time, energy, and the production quality will also be enhanced. The use of this principle is evident from the metal treatment processes of direct metal deposition vs the metal casting [6].

6. Conservation of complexity of the design:
This principle of sustainable production deals with the conservation of the complex designs adopted for products manufacturing in terms of

the benefits obtained from the investments which are done for the complex designs. Processes with the complexities are usually function of the material expenditure, energy expenditure, and time consumption. The designers should take the benefits from these complexities in the form of counter production or reuse, e.g., silicon chips in modern computer systems installed with more complexities tend to maximize the output performance in terms of time saving, and other certain factors as compared to the previous versions of Intel with relatively low complexities, but the output performance was rather poor than the former.

7. Focus on the durability rather than the immortality:
The designers should take the concern for the goal of product durability rather than immortality of the product to reduce the threat and risks to humans and to the environment. Products which are designed based on the immortality can put serious threats to the environment in the form of solid wastes, nonbiodegradable substances, etc. So, the designs based on the limited and targeted lifetime needs to be designed. It is necessary to keep an eye on the stagewise maintenance operations in the process, and assembly to avoid any pre lifetime failure of the assembly and to avoid the following disposal problems as well. Keeping in view this limited and targeted lifetime will ultimately minimize the extra need of resources in the form of materials input, energy, and time. Thus, the process will lead towards the sustainable production. As an example, the use of disposable diapers has put a commendable amount of contribution to the environmental stress as they are composed of the materials which are poor in terms of the biodegradability and recycling of the municipal solid wastes. The solution to put an end to this environmental problem could be the use of starch and water-based materials which are easily self-soluble and biodegradable. The production of plastic and fibers-based components from petrochemical based polyacrylic acid need to replace by biological based polylactic acid which are kind to the environment as well as leads towards the sustainable production.

8. Meet the needs and minimize the excess:
To check the ability of a process to move fast, easily, and flexibly at the design stage is very important. The designer should take the necessary steps for the disposal, and monitoring of the process at each stage of the process to overcome the extra production, which leads to unsustainability in terms of the materials and energy usage. In short, the design and manufacturing engineers should incorporate the design ideas to minimize the expenditures required for the resources. Production of the same amount

of quantity for various levels of need will ultimately lead to environmental stresses in the form of harmful byproducts in enhanced amount. The purification or disinfection process of water by the method of chlorination could be the example of this case. The buildings available at the closest to the purification plant will be provided with the excess amount of chlorine product than those available at far from the purification plant and thus moderate amount or the amount according to the exact need will be provided. This is because this disinfecting product of chlorine degrades with respect to time. The problem can be overcome by the installation of the proper control systems throughout the distribution system. The results could be the reduction of environmental stresses and risks to human health in the form of trihalomethanes, the byproduct of disinfecting agents of chlorine [7].

9. Minimize the materials diversity in multicomponent assemblies:
Products having extensive use in daily life such as automobiles, computer systems, and packaging of food items are made up of diverse assemblies including woods, metals, composites, glasses, and plastics. Each individual class of materials has their own negatives in the form of unkindness to the environment and human's life. The typical plastic material contains various dyes, fire retardants, plasticizers, stabilizers, and various hazardous chemical additives. These extensive amounts of components used in the products could create stress to both the environment and living beings present on the planet. The design and manufacturing engineers should consider the end-of-life criteria for the disposition, recycle, or reuse of the product. This is possible by adopting the design ideas for the assemblies which are easy to disassemble. The use of polymers as an alternative for the hazardous chemical additives can give some green effects, i.e., modifying the polymers properties for the required purpose can have beneficial effects in terms of environmental benign processes. The design and manufacturing engineers should try to build up the product prototypes based the single material use, e.g., the use of metallocene polyolefins for the product building. The analogues design ideas on molecular level could be explained by cascading reactions vs multistage reactions which would require some extra chemicals for providing energy in the form of ignition [8].

10. How to reuse the consumed materials and energy:
The traditional design and manufacturing techniques tend to use the materials and energy at input of the process, some of the resources used for operating the process and usually large fraction of these materials or energy wastes in the form of heat energy or, say industrial wastes. So,

design ideas of how to recycle these resources should be incorporated into modern design ideas. Recycling these resources in an efficient way could reduce the additive expenditures required for the resources in the form of materials, and energy which would run the subcomponents of the operating unit. On industrial scale, the energy produced because of the large amounts of exothermic reactions usually wastes in significant amount in the form of heat generated in the process, or some other forms of energy. This much amount of energy needs to be recycled to reuse as the input for the process or to run some additional components used in the operating unit. This fundamental principle of green engineering could also be applied to chemicals industry by storing the byproducts obtained because of the many types of reactions, e.g., the byproducts obtained during the purification process of the chemically synthesized products. The innovative designs of cogeneration to produce electricity from industrial wastage energy could be the typical example.

11. Designing commercial "afterlife" ideas:
Sometimes the market value of the valuable product become obsolete due to the new advances and the introduction of the new same purpose products which are more sustainable, efficient, and stylistic in looking, but the working capability of the product is still alive. To cope with the competitive status of the market, the designers should develop the ideas to use the existing components in the upcoming next generation of the product rather than to invest new costly resources in the form of materials, and energy. By incorporating the fundamental "afterlife" design idea in the very initial stage of the process could give some fruitful results in terms of the reinvesting the available resources. This principle of green engineering can give some good results for the introduction of new generations of the mobile phones, computers, and laptops as every new year the new generation of the former mentioned products comes into market, creating the competitive environment for the designers and manufacturers. Xerox instrument used approximately 90% of the components which could be remanufactured. The use of old building materials in a new house could also be a typical example.

12. Use renewable resources rather than depleting resources:
The designers should try to study the ultimate origin of the input material and energy required in a product operating process, because this can have far and wide effects on the product sustainability. The use of pure substances at each input stage requires certain extraction, separation, and

purification processes, so the use of new depleting materials at every input stage can produce certain environmental rifts. The use of biological renewable resources in this aspect is preferred. The application area of this principle of green engineering can be seen in wastewater treatment through natural ecosystem, biomass feedstock's recovery, and biological based plastics rather than additives-based plastics [9].

The principles discussed above are not merely a set of goals, but a fundamental set of design and manufacturing methodologies to ultimately achieve the desired goal of green engineering, green technology, clean technology, and sustainability.

6.3 Benefits of Green Engineering

Here are several benefits that a company may experience after implementing green engineering practices:

- *Enhances business practices*
Green engineers may visit your facility, identify problematic or inefficient operations and improve them. This process may enhance business practices by eliminating improper production methods and helping you adheres to federal or state environmental regulations. Improved manufacturing or shipping processes may also improve your product quality and earn you more sales.

- *Improves a company's reputation*
Green engineering may improve a company's reputation by showing consumers it cares about the environment. You may print press releases highlighting your changes and enter new customer markets from these actions. You may even identify new customers who may work with you after you change to green energy sources and techniques.

- *Minimizes energy or production waste*
An inefficient operation may cost company money, but green engineers can help solve this problem. They identify where you're wasting energy or production efficiency, identify potential solutions, install innovative methods and provide maintenance for your manufacturing machines. Green engineers may install alternative energy sources, reduce your production steps, enhance sustainability by identifying better construction materials and reduce your energy bills.

• *Provides tax incentives*
Federal and state governments often provide tax credits and rebates for companies and individuals who use more sustainable processes. Green engineers can help you earn these credits by installing more efficient energy and operating systems in your facility. These rebates may include large tax breaks that help offset certain taxes and increase profits by helping executive personnel invest this money back into their company.

• *Helps the global environment*
Improved energy efficiency, more sustainable production methods, longer-lasting products and streamlined manufacturing may provide real benefits to communities. Decreasing your carbon or waste profile can decrease your area's pollution output. Even minor changes may benefit your area by providing a better environmental example for others, such as your competitors and your customers.

6.4 Promoting Green Engineering Through Green Chemistry

The decision of chemical engineers may bring fruitful results for the manufacturing engineering by making decisions for the products and processes they made. The properties, physical or chemical, of the substances or chemicals identified by the chemical engineers may highlights the use of a suitable machine, or say component in the form of reactor, boiler, condenser, etc. that to be used in the process. The life of the design and manufacturing engineers is simplified by the chemists in terms of the minimization or elimination of the use and generation of the hazardous substances. The integral use of green engineering and green chemistry is requiring for the sustainable production. Green chemistry ensures the safety of the operating process by the selection of safe substances, and in terms of the selection of the reagents, solvents and feedstock used [10].

Green chemistry applications:

a) Feedstocks:
Choosing the feedstock for certain manufacturing process, the factors of toxicity, source of the feedstock, and effects on the environments should be of the prime interest to be taken under the consideration. The researchers are trying to develop the biological renewable feedstock rather than the petroleum based feedstock. Biological based feed stocks include the use of

agricultural biomass, carbon dioxide, chitin, and waste by products. The efficiency of the two sources observed until now gives favor toward the use of petroleum-based feedstock resources.

b) Reagents:
Most of the chemical reagents used in industry are corrosive, toxic, hazardous, and unkind to the environment and public as well. The oxidation chemistry is a useful substitution to the above problem to carry out the chemical processes in a safe manner. The emerging product of the oxidation chemistry is alcohol. Sheldon and cooperators have designed systems based on clean technology to convert alcohols into aldehydes and ketones. Other system Sheldon and the coworkers combine ruthenium catalyst with TEMPO (a chemical substance) under aerobic media to oxidize a vast amount of alcohol and the only side product in the process is water.

Phosgene, an intense toxic substance is widely used as a chemical reagent in the industry. The possible substitution for the Phosgene is Dimethyl carbonate (DMC), which is widely used for the methylation purposes in place of methyl halides and phosgene chemicals. Dimethyl carbonate (DMC) turns out to be prepared in a no phosgene media from methanol and carbon dioxide. The processes associated with the Dimethyl carbonate (DMC) are usually carried out at high temperatures and pressures, but the processes are safe and kind to the environment leading to sustainable production.

c) Solvents:
Solvents are widely used in industry to carry out certain reactions. Many of these solvents are organic volatile substances, dangerous air pollutants, flammable, explosive, and toxic in nature. These associated factors with the solvents used on the industrial scale have turned out the attention of chemists to avoid the stress on environment and the risks to human life. The use of supercritical fluids, water, and ionic liquids could be the possible and useful substitution the traditional solvents used in industry.

6.5 Sustainability and Green Engineering Innovations That Might Just Change the World

The world must change if it is to survive. There has never been a greater interest and push for the sustainable products and technologies that there is today. The world has reached to a critical point regarding the climate change, and many innovators, businesses, and designers are stepping up to the level to build a green future [11].

Following in the discussion are some incredible examples of sustainable innovation that could positively change the world as we know it, and they pave the way for a more sustainable way of living on the face of this earth.

1. The smog free project:
The smog free project is a very long-term campaign for clean and green air in which Dann Roosegaarde and his teammates have created the world's first smog vacuum cleaner. This 7-meter-tall smog free tower utilizes the patented positive ionization technology for production of smog free air in public areas like parks, hospitals, and resorts, allowing the people to use the fresh and clean air free of cost. This tower is equipped with the environmentally friendly technology and cleans 30,000 m3 of air per hour. The resource require for the operation is small amount of electricity which is also produced by green technology for this specific tower. The function and working of smog free tower have been legalized by the results compiled by the Eindhoven University of Technology.

2. Zephyr solar balloon:
Zephyr is a photovoltaic balloon and environment friendly generator created by Karen Assaraf, Julie Dautel, and Cedric Tomissi. This aerial platform takes the shape of helium balloon. These solar balloons are operated individually or in the form of aerial network. These solar balloons can be installed within an hour, can fly up to 30 days at a time, they are able to withstand winds having speed of 43mph, and can fly at great altitudes. These balloons can be used for a vast range of sustainable activities. They are incorporated in various scientific missions observing the wildlife, to measure the air quality, and can be used to enhance the coordination during any natural disaster. This solar balloon for the first time reported in France.

The modern day technological team needs to take serious concern for introducing these solar balloons on large worldwide [12].

3. The Green Building Initiative:
The green building initiative (GBI) is a nonprofit organization introduced to accelerating the adopting of building practices that usually results is resource efficient, healthier and buildings which are environmentally sustainable. The strategy to achieve the desired goals involves education and promotion on the use of GBI's Green Globes certification program for the commercial buildings and GBI's Guiding Principles Compliance and third party assessment program for federal building sustainability requirements.

4. B-Droid:
B-Droid is one of the several modern-day efforts to crate robotic bees that can pollinate crops as effectively as their organic alternatives. B-Droid's aim is to help boost the population of natural bees by growing the low nutritional and high labor tasks of pollination to these robotic bees [13].

5. Groasis Waterbox:
The Groasis water box, created by the Dutch flower exporter, Pieter Hoff, is a device that makes growing crops possible and resources efficient even in

the deserts. It consists of an intelligent bucket made from the paper, which is recyclable, and helps in seed germination, and incubate the tree plants. It usually requires 90% less than traditional growing methods and can be used in some of the extreme climate regions [14].

6. AirCarbon:

AirCarbon (2020) was developed by Newlight technologies and has already won many awards for its sustainability. This AirCarbon is made from the carbon emissions that are usually just released into the atmosphere. It is a verified and legal carbon-negative material, i.e., every step of its production and use is fully sustainable and green. It is the best alternative to other synthetic materials as it is not made from the non-degradable usual plastics.

7. Solar Glass:

The discovery of the solar glasses could change the world the way we create the modern-day homes and commercial buildings. Researchers at the University of Michigan are trying to develop the solar glass, being a sustainable and green product that has generated a lot of talk in the modern-day world. The glass would be able to capture and store the solar energy, if used as an alternative for the conventional brick's buildings.

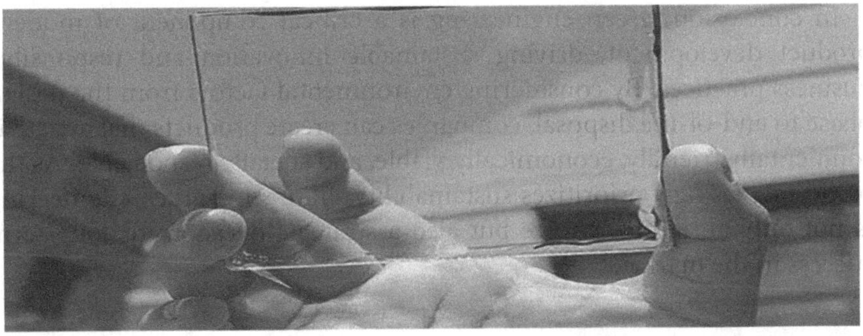

6.6 Conclusion

Green engineering has emerged as a crucial aspect of product development in response to the growing concern for environmental sustainability. This paper has provided insights into the principles and practices of green engineering, emphasizing the need for integrating environmental considerations throughout the product lifecycle. By adopting green engineering strategies, companies can not only reduce their environmental footprint but also gain a competitive edge in the market.

The implementation of green engineering practices offers several benefits. Firstly, it helps in reducing resource consumption and waste generation, leading to cost savings through improved efficiency. By optimizing material usage and energy consumption, companies can minimize production costs and enhance profitability. Additionally, green engineering enables companies to comply with environmental regulations and standards, mitigating the risk of legal penalties and reputational damage.

Furthermore, embracing green engineering can improve brand reputation and customer loyalty. With the increasing awareness and demand for sustainable products, consumers are actively seeking environmentally friendly options. By demonstrating a commitment to sustainability through green engineering practices, companies can attract eco-conscious consumers and differentiate themselves from competitors.

To effectively incorporate green engineering in product development, collaboration and interdisciplinary approaches are essential. Design teams need to work closely with environmental experts, suppliers, and stakeholders to identify and implement sustainable solutions. Life cycle assessment (LCA) and eco-design methodologies can be utilized to evaluate the environmental impacts of products and guide decision-making processes.

In conclusion, green engineering is a critical component of modern product development, driving sustainable innovation and responsible business practices. By considering environmental factors from the design phase to end-of-life disposal, companies can create products that are environmentally friendly, economically viable, and socially responsible. As the world increasingly prioritizes sustainability, embracing green engineering is not only an ethical choice but also a strategic decision for long-term success in the market.

References

1. Mulvihill, M.J., Beach, E.S., Zimmerman, J.B., Anastas, P.T., Green chemistry and green engineering: a framework for sustainable technology development. *Annu. Rev. Environ. Resour.*, 36, 271–293, 2011.
2. García-Serna, J., Pérez-Barrigón, L., Cocero, M.J., New trends for design towards sustainability in chemical engineering: Green engineering. *Chem. Eng. J.*, 133, 1–3, 7–30, 2007.
3. Kirchhoff, M.M., Promoting green engineering through green chemistry. *Environ. Sci. Technol.*, 37, 23, 5349–5353, 2003.
4. Kavitha, V. and Mohanraj, S., Green engineering principles for global water quality monitoring using IoT. *Int. J. Environ. Sustain. Dev.*, 18, 1, 120–129, 2019.
5. Ardila-Fierro, K.J. and Hernández, J.G., Sustainability assessment of mechanochemistry by using the twelve principles of green chemistry. *ChemSusChem*, 14, 10, 2145–2162, 2021.
6. Mihelcic, J.R. and Zimmerman, J.B., *Environmental engineering: Fundamentals, sustainability, design*, John wiley & sons, 2021.
7. Erythropel, H.C., *et al.*, The Green ChemisTREE: 20 years after taking root with the 12 principles. *Green Chem.*, 20, 9, 1929–1961, 2018.
8. Arbabzadeh, M., Johnson, J.X., Keoleian, G.A., Rasmussen, P.G., Thompson, L.T., Twelve principles for green energy storage in grid applications. *Environ. Sci. Technol.*, 50, 2, 1046–1055, 2016.
9. Segars, J.W., Bradfield, S.L., Wright, J.J., Realff, M.J., EcoWorx, green engineering principles in practice. *Environ. Sci. Technol.*, 37, 23, 5269–5277, 2003.
10. Zimmerman, J.B., Anastas, P.T., Erythropel, H.C., Leitner, W., Designing for a green chemistry future. *Science*, 367, 6476, 397–400, 2020.
11. Xie, X., Huo, J., Zou, H., Green process innovation, green product innovation, and corporate financial performance: A content analysis method. *J. Bus. Res.*, 101, 697–706, 2019.

12. Westbrook, T. and Rostvik, H., Solar energy and battery development for unmanned aerial vehicles: Leading to increased proliferation? *Salus J.*, 6, 2, 59–77, 2018.
13. Antolin-Lopez, R., Garcia-de-Frutos, N., Gálvez-Rodríguez, M.M., Tarifa-Fernandez, J., Saez-Martin, A., Emotional learning in business & sustainability education: A classroom methodology proposal, in: *EDULEARN18 Proceedings*, IATED, pp. 690–695, 2018.
14. Robles, J.L.M., *El control de la supervivencia en plantaciones agroforestales con waterboxx mediante vehículos aéreos no tripulados e información multiespectral*, Universidad de Valladolid, 2016, PhD Thesis.

7

Green Policies in Education: Fostering Environmental Stewardship and Sustainable Practices

Mukhtiar Singh*, Maninder Singh, Jasvinder Singh, Harjit Singh and Mandeep Singh

Lovely Professional University, Jalandhar - Delhi, Phagwara, Punjab, India

Abstract

This research study investigates the impact of green policies in education on fostering environmental stewardship and promoting sustainable practices among students. The study synthesizes existing research and explores key insights, implications, and recommendations for policymakers, educators, and researchers. Through a comprehensive review of literature, the study reveals that green policies integrated into educational systems play a crucial role in developing students' knowledge, awareness, and sense of responsibility toward environmental issues. The findings emphasize the effectiveness of integrating environmental education across the curriculum, enabling students to comprehend the interconnectedness of environmental challenges with various subjects. Experiential and outdoor learning experiences emerge as significant components of green policies, fostering hands-on engagement and nurturing students' connection with nature. The study underscores the importance of providing adequate teacher training and professional development programs to equip educators with the necessary knowledge and skills to integrate environmental concepts effectively. Furthermore, the study highlights the significance of evaluating the outcomes and impact of green policies and educational initiatives. It emphasizes the need for robust evaluation frameworks and assessment methods to measure the effectiveness of environmental education programs, ensuring continuous improvement and accountability. The study also emphasizes the value of community engagement and partnerships, as well as the importance of addressing equity and inclusion in environmental education. Culturally responsive approaches and consideration of environmental justice

**Corresponding author*: mukhtiar.16431@lpu.co.in

Chandan Deep Singh, Talwinder Singh and Davinder Singh (eds.) Digitization and Manufacturing Performance: An Environmental Perspective, (167–210) © 2025 Scrivener Publishing LLC

issues are crucial for ensuring that environmental education reaches all students, regardless of their backgrounds. Based on the research findings, the study provides practical recommendations for policymakers and educators to enhance the implementation of green policies in education. These recommendations include curriculum integration, promotion of experiential learning, provision of teacher training and professional development, improvement of school infrastructure, fostering community partnerships, and addressing equity and inclusion. In conclusion, this research study underscores the significance of green policies in education for fostering environmental stewardship, promoting sustainable practices, and shaping a more environmentally conscious society. By implementing the insights and recommendations derived from this study, policymakers, educators, and researchers can contribute to the development of effective and impactful environmental education initiatives.

Keywords: Green policies, environmental education, sustainability, experiential learning, teacher training, community engagement, equity, evaluation frameworks

7.1 Introduction

Over the past few decades, there has been growing concern about environmental sustainability and the need for a transition to a greener and more sustainable future. Education plays a crucial role in shaping the attitudes, knowledge, and skills of individuals, and as a result, it has become an important arena for promoting green policies and practices. One of the key drivers behind green policies in education is the recognition of the environmental challenges facing the world. Issues, such as climate change, deforestation, loss of biodiversity, and pollution, have gained global attention, and there is a growing consensus that urgent action is needed to address these challenges. Education is seen as a powerful tool to raise awareness about these issues, foster sustainable behavior, and equip individuals with the knowledge and skills needed to contribute to sustainable development. The United Nations has been at the forefront of promoting green policies in education through its global initiatives. The most notable of these is the United Nations Decade of Education for Sustainable Development (2005–2014), which aimed to integrate sustainable development principles into all aspects of education. This initiative emphasized the importance of integrating environmental, social, and economic dimensions of sustainability into the curriculum, pedagogy, and management of educational institutions. Following the UN Decade, many countries and educational institutions have developed their own green policies and programs. These policies typically include integrating sustainability into the curriculum across various disciplines, promoting environmental awareness and responsibility, adopting sustainable practices within educational institutions, and engaging with the local community and stakeholders in sustainability initiatives. The Decade of Education for Sustainable Development emphasized the need to integrate sustainability principles into the curriculum, pedagogy, and management of educational institutions. It called for a shift in the way education is delivered, moving away from traditional disciplinary boundaries and toward interdisciplinary and transdisciplinary approaches. The aim was to foster a deep understanding of sustainability issues and encourage critical thinking, problem-solving, and active citizenship among learners. To achieve these goals, the initiative highlighted the importance of teacher training and capacity building. It called for the development of educational materials, tools, and resources that promote sustainable development and provide guidance to educators. The idea was to empower teachers to integrate sustainability concepts into their teaching practices and create innovative and engaging learning experiences. The Decade of Education for Sustainable Development also emphasized the role of educational institutions as models of sustainability. It encouraged schools, universities, and other educational organizations to adopt sustainable practices in their operations and management. This involved initiatives such as reducing energy and water consumption, promoting waste reduction and recycling, implementing sustainable transportation options, and adopting green building designs. Furthermore, the initiative recognized the significance of partnerships and collaboration among various stakeholders. It called for the engagement of governments, non-governmental organizations (NGOs), businesses, communities, and other sectors in supporting and promoting green policies in education. Partnerships were seen as a means to mobilize resources, share best practices, and create a collective effort toward sustainability. Throughout the decade, numerous activities and events were organized at local, national, and international levels to promote the goals of the initiative. These included conferences, workshops, training programs, and campaigns to raise awareness about sustainability and showcase successful examples of green policies in education. The Decade of Education for Sustainable Development concluded in 2014, but its legacy continues to influence education systems and policies worldwide. It laid the foundation for ongoing efforts to integrate sustainability into education and inspired subsequent initiatives such as the Global Action Program on Education for Sustainable Development (2015-2019) and the Sustainable Development Goals (SDGs), particularly SDG 4, which focuses on quality education and SDG 13, which addresses climate action. Green policies in education often involve curriculum reform to incorporate sustainability concepts and principles into different subjects.

This can include topics, such as climate change, renewable energy, waste management, biodiversity conservation, and sustainable agriculture. The goal is to provide students with a comprehensive understanding of the environmental challenges and potential solutions. Educational institutions have also taken steps to become more sustainable themselves by implementing eco-friendly practices. This can involve energy conservation, waste reduction and recycling, water conservation, sustainable transportation, and green building design. By modeling sustainable practices, educational institutions aim to create a conducive environment for learning and promote a culture of sustainability among students and staff. Furthermore, green policies in education emphasize the importance of experiential and outdoor learning. This approach encourages students to engage directly with the environment, fostering a sense of connection and responsibility toward nature. Outdoor activities, field trips, and environmental projects provide opportunities for students to apply their knowledge and develop practical skills in real-world contexts. The implementation of green policies in education is often supported by partnerships and collaborations between educational institutions, governments, non-governmental organizations (NGOs), and local communities. These partnerships can provide additional resources, expertise, and networking opportunities to enhance the impact of green initiatives.

7.1.1 Importance and Relevance of Green Policies in the Education Sector

Green policies in the education sector hold immense importance and relevance due to several key reasons. These policies play a crucial role in addressing environmental challenges, fostering sustainability awareness and behavior, equipping individuals with the knowledge and skills for sustainable development, and creating a greener future. Here are some references that highlight the significance of green policies in education:

- **United Nations Educational, Scientific and Cultural Organization (UNESCO):**

UNESCO emphasizes the importance of integrating sustainable development principles into education to address global challenges. They advocate for education systems that promote environmental stewardship, social responsibility, and economic viability.

• **United Nations Environment Programme (UNEP):**
UNEP recognizes the critical role of education in promoting sustainable development and the transition to a green economy. They emphasize the need to incorporate environmental education into curricula to empower individuals to make informed decisions for a sustainable future. Reference: UNEP

• **International Union for Conservation of Nature (IUCN):**
IUCN stresses the importance of integrating environmental education into school curricula to build a generation of environmentally conscious citizens. They highlight the role of education in raising awareness, developing critical thinking, and inspiring action for biodiversity conservation and sustainable development.

• **Worldwatch Institute:**
The Worldwatch Institute highlights the need for green policies in education to ensure that future generations are equipped with the knowledge and skills to tackle environmental challenges. They argue that education is a key driver of behavior change and societal transformation toward sustainability.

• **European Commission:**
The European Commission recognizes the importance of green policies in education for achieving the objectives of the European Green Deal. They emphasize the integration of sustainability in education and training systems to promote green skills, knowledge, and innovation.

These references highlight the significance of green policies in education, as recognized by international organizations and initiatives. They underscore the role of education in addressing environmental challenges, promoting sustainable behavior, and equipping individuals with the necessary skills for a sustainable future.

7.1.2 Objectives of the Research

Following are the main research objectives of the study:

- Assessing the Effectiveness of Green Policies: Research aims to evaluate the effectiveness of green policies implemented in the education sector. This involves examining the extent to which these policies achieve their intended goals, such as promoting sustainability awareness, changing behaviors, and integrating sustainability principles into education systems.

- Understanding the Impacts on Students and Educators: Research seeks to understand the impacts of green policies on students and educators. This includes examining changes in knowledge, attitudes, and behaviors related to sustainability among students. It also explores the experiences and challenges faced by educators in implementing green policies and integrating sustainability into their teaching practices.
- Examining Curriculum Integration: Research focuses on investigating how sustainability concepts and principles are integrated into the curriculum across different subjects and educational levels. It aims to assess the depth and breadth of sustainability education, identify gaps or barriers, and propose strategies for effective curriculum integration.
- Evaluating Institutional Practices: Research examines the adoption and implementation of sustainable practices within educational institutions. This involves assessing the extent to which institutions have adopted environmentally friendly practices such as energy and resource conservation, waste management, sustainable transportation, and green building designs. The objective is to evaluate the impact of these practices and identify areas for improvement.
- Identifying Best Practices and Lessons Learned: Research aims to identify best practices and lessons learned from the implementation of green policies in education. This involves examining successful case studies and initiatives that have effectively integrated sustainability into education systems. By identifying and sharing these best practices, research aims to inform policy development, program implementation, and institutional change.
- Informing Policy Development: Research plays a crucial role in informing the development of green policies in education. By providing evidence-based insights, research helps policymakers understand the benefits, challenges, and potential strategies for promoting sustainability in education. It supports the formulation of policies that are grounded in research findings and align with broader sustainability goals.
- Contributing to Knowledge and Theory: Research on green policies in education contributes to the broader knowledge base and theory related to sustainability education. It helps

expand our understanding of effective approaches, methodologies, and pedagogies for integrating sustainability into education systems. Research findings contribute to academic scholarship, theoretical frameworks, and methodologies for future research and practice.

7.1.3 Significance of the Study

The study on Green Policies in Education holds significant importance due to several reasons. By examining the implementation and impacts of green policies in education, this research contributes to the broader understanding of sustainability education and its role in fostering environmental consciousness, behavior change, and sustainable development. Here are some key significance of studying Green Policies in Education:

- Informing Policy and Decision-Making: Research on green policies in education provides valuable insights that inform the development and refinement of policies and strategies related to sustainability education. Policymakers can utilize research findings to make evidence-based decisions and design effective policies that align with sustainability goals.
- Enhancing Educational Practices: The study of green policies in education helps identify best practices and effective approaches to integrate sustainability into educational practices. It informs educators and educational institutions about successful strategies, methodologies, and pedagogies for promoting environmental awareness and sustainability literacy among students.
- Assessing Effectiveness: Research allows for the assessment of the effectiveness of green policies in education. By evaluating the outcomes and impacts of these policies, researchers can determine the extent to which they have achieved their intended goals, such as raising awareness, changing behaviors, and integrating sustainability principles into educational systems.
- Promoting Sustainability Awareness and Behavior: Understanding the impacts of green policies in education helps in identifying the most effective approaches to fostering sustainability awareness and promoting environmentally responsible behaviors among students. This knowledge

contributes to the development of educational interventions and programs that cultivate a culture of sustainability.

- Enhancing Curriculum Integration: Research on green policies in education explores the integration of sustainability concepts and principles into the curriculum across various subjects and educational levels. It helps identify gaps, challenges, and opportunities for further curriculum development and integration of sustainability education, ensuring a comprehensive and holistic approach to sustainability across disciplines.
- Supporting Institutional Change: The study of green policies in education contributes to institutional change within educational settings. It provides insights into the adoption and implementation of sustainable practices within educational institutions, encouraging the transition toward environmentally friendly operations and management.
- Contributing to Global Sustainability Goals: Research on green policies in education aligns with global sustainability goals, such as the United Nations Sustainable Development Goals (SDGs). By promoting sustainable development through education, research in this area directly contributes to the achievement of these global goals.
- Empowering Future Generations: The significance of studying green policies in education lies in its potential to equip future generations with the knowledge, skills, and values necessary to address environmental challenges and contribute to a sustainable future. By understanding effective strategies and practices, researchers can support the empowerment of students to become active agents of change.

In conclusion, the study of green policies in education holds immense significance as it informs policy and decision-making, enhances educational practices, assesses effectiveness, promotes sustainability awareness and behavior, enhances curriculum integration, supports institutional change, contributes to global sustainability goals, and empowers future generations. Through research in this field, we can create more effective and impactful green policies in education that contribute to a more sustainable and environmentally conscious society.

7.2 Theoretical Framework

There are several relevant theories and concepts related to green policies in education that provide frameworks for understanding and implementing sustainability education. These theories and concepts help guide the development and implementation of green policies and practices in educational settings. Here is a review of some of these theories and concepts.

- **Education for Sustainable Development (ESD):**
ESD is a concept that emphasizes the integration of sustainable development principles into all aspects of education. It promotes an interdisciplinary approach to learning and aims to equip learners with the knowledge, skills, values, and attitudes necessary to contribute to a sustainable future. ESD recognizes the interconnectedness of environmental, social, and economic dimensions of sustainability [1].

- **Transformative Learning:**
Transformative learning theory, developed by Jack Mezirow, suggests that education has the power to transform individuals' perspectives, values, and behaviors. In the context of green policies in education, transformative learning involves challenging and changing learners' deeply held beliefs and assumptions about sustainability issues, leading to personal and societal transformation [2].

- **Environmental Education:**
Environmental education focuses on raising awareness and understanding of environmental issues and promoting responsible behavior toward the environment. It encompasses concepts such as ecological literacy, environmental ethics, and nature-based learning experiences. Environmental education often serves as a foundation for green policies in education [3].

7.2.1 Place-Based Education

Place-based education emphasizes the importance of connecting students to their local environments and communities. It encourages learning experiences that are rooted in the local context, fostering a sense of place, environmental stewardship, and community engagement. Place-based education can be integrated into green policies to enhance students' understanding of and connection to their local ecosystems [4].

• Systems Thinking:

Systems thinking are an approach that recognizes the interconnectedness and complexity of natural and social systems. It encourages understanding the relationships and interdependencies within systems and the consequences of actions. Green policies in education can incorporate systems thinking to help students analyse and address sustainability challenges holistically [5].

• Ecological Citizenship:

Ecological citizenship refers to the notion that individuals have rights, responsibilities, and duties as citizens of the Earth. It encompasses the idea that individuals should actively participate in environmental decision-making and contribute to the well-being of the planet. Green policies in education can foster ecological citizenship by empowering students to become engaged and responsible environmental stewards [6].

These theories and concepts provide frameworks for understanding the principles, approaches, and goals of green policies in education. They inform the design and implementation of sustainability education initiatives, guiding educators, policymakers, and researchers in their efforts to promote environmental awareness, behavior change, and sustainable development within educational settings. When examining policy implementation theories in the educational context, several theoretical frameworks are commonly used to understand the complexities and challenges associated with policy implementation. Here are a few policy implementation theories in the educational context:

• Advocacy Coalition Framework (ACF):

The Advocacy Coalition Framework focuses on the interactions between advocacy coalitions, policymakers, and policy implementation processes. It examines the role of different actors, their beliefs, resources, and interactions in shaping policy implementation outcomes. Research using the ACF in education has explored issues such as the implementation of inclusive education policies [7].

• Institutional Theory:

Institutional theory examines the role of formal and informal institutions in shaping policy implementation processes. It emphasizes the importance of organizational structures, norms, and routines in shaping how policies are interpreted and implemented in educational settings. Research utilizing institutional theory has explored the implementation of various education policies, such as curriculum reforms and accountability policies [8].

• **Diffusion of Innovations Theory:**
Diffusion of Innovations Theory focuses on how new ideas, practices, or policies spread and are adopted within a social system. It explores the factors that influence the adoption and implementation of educational policies and innovations, including the characteristics of the policy itself, the adopters, and the context in which implementation takes place [9].

• **Policy Networks:**
Policy network theory examines the relationships and interactions among various actors involved in policy implementation, such as policymakers, educators, administrators, and community stakeholders. It explores how these actors collaborate, negotiate, and influence policy implementation processes in education [10].

The relationship between green policies and sustainable development goals (SDGs) is closely interconnected and mutually reinforcing. Green policies are instrumental in achieving the environmental dimension of sustainable development, which is encapsulated in several SDGs. Here is an analysis of the relationship between green policies and sustainable development goals:

7.2.2 Addressing Environmental Challenges

Green policies aim to tackle pressing environmental challenges such as climate change, biodiversity loss, pollution, and resource depletion. These policies encompass a range of measures, including renewable energy promotion, energy efficiency regulations, waste management strategies, sustainable agriculture practices, and conservation initiatives. By implementing green policies, countries can make significant progress toward achieving specific environmental targets embedded within the SDGs.

• **Contribution to Multiple SDGs:**
Green policies contribute to the achievement of multiple SDGs, demonstrating their cross-cutting nature. For instance:

- Green energy policies and investments in renewable energy contribute to SDG 7 (Affordable and Clean Energy).
  - ➢ Sustainable agriculture practices and land-use policies support SDG 2 (Zero Hunger) and SDG 15 (Life on Land).
  - ➢ Policies targeting waste management and circular economy principles align with SDG 12 (Responsible Consumption and Production).

- Efforts to mitigate climate change through emissions reduction contribute to SDG 13 (Climate Action).
- The integration of green policies across different sectors can generate synergies and maximize the positive impacts on sustainable development.

• **Policy Coherence:**
Green policies promote policy coherence by considering the interlinkages between environmental, social, and economic dimensions of sustainable development. By aligning with the SDGs, green policies ensure a comprehensive and integrated approach to addressing global challenges. Policy coherence allows for the identification of potential trade-offs or complementarities between different goals and facilitates coordinated action across sectors and levels of governance.

• **Implementation Challenges:**
Implementing green policies to achieve the SDGs poses challenges such as balancing environmental priorities with social and economic considerations, ensuring equitable access to green technologies and practices, and addressing the varying capacities of different countries. Additionally, coordination among diverse stakeholders, including governments, businesses, civil society, and communities, is crucial for effective policy implementation and to harness collective efforts toward sustainable development.

The relationship between green policies and sustainable development goals is dynamic and evolving, driven by the need for transformative changes to achieve environmental sustainability. By integrating green policies within the broader framework of the SDGs, countries can work toward a more inclusive, resilient, and environmentally sustainable future.

7.3 Policy Development and Implementation

The policy development process in the education sector involves a series of steps that encompass research, analysis, stakeholder engagement, decision-making, implementation, and evaluation. Here is an overview of the policy development processes in the education sector, along with some research references:

• **Problem Identification and Agenda Setting:**
The policy development process begins with identifying key issues and challenges in the education sector. This stage involves conducting research,

analysing data, and consulting stakeholders to determine the problems that require policy attention. The agenda-setting process involves prioritizing issues based on their significance, urgency, and feasibility [11].

- **Policy Formulation:**
Policy formulation involves the development of policy proposals and options to address the identified issues. This stage requires research, analysis of evidence, and the involvement of experts to design policies that are informed by best practices and evidence-based approaches [12].

- **Stakeholder Engagement and Consultation:**
Engaging stakeholders, including teachers, students, parents, education professionals, community organizations, and policymakers, is crucial in the policy development process. Stakeholder engagement allows for diverse perspectives, expertise, and experiences to be considered, leading to more inclusive and effective policies [13].

- **Decision Making and Adoption:**
Once policy proposals are developed, decision-making processes take place to select the preferred policy options. Decision-makers, such as government officials, education authorities, and legislative bodies, review and assess the proposals, considering political, economic, social, and educational factors. The chosen policy options are then adopted as official policies [14].

7.3.1 Policy Implementation

Policy implementation involves translating the adopted policies into action. This stage includes developing implementation plans, allocating resources, establishing monitoring and evaluation mechanisms, and ensuring effective communication and coordination among stakeholders. Implementation also involves addressing potential challenges and barriers that may arise [15].

- **Policy Evaluation:**
Policy evaluation assesses the effectiveness, efficiency, and impact of the implemented policies. Evaluation helps determine whether the policies have achieved their intended goals, identifies areas for improvement, and informs future policy decisions. Various evaluation methods, such as qualitative and quantitative research, monitoring, and data analysis, are used to evaluate policies in the education sector [16].

Analysis of international and national policies promoting environmental sustainability in education reveals a growing recognition of the importance of integrating sustainability into educational systems. These policies aim to enhance environmental awareness, knowledge, and skills among learners, foster a sense of responsibility toward the environment, and promote sustainable practices. Here is an analysis of international and national policies promoting environmental sustainability in education:

- **International Policies:**

a. United Nations Decade of Education for Sustainable Development (2005-2014):
This initiative aimed to integrate sustainable development principles into education systems worldwide. It emphasized the need to incorporate environmental, social, and economic dimensions of sustainability into curriculum, pedagogy, and management of educational institutions.
b. United Nations Sustainable Development Goals (SDGs):
The SDGs, particularly SDG 4 (Quality Education) and SDG 13 (Climate Action), highlight the importance of environmental sustainability in education. The SDGs call for inclusive and equitable quality education that promotes sustainable development and addresses climate change.

- **National Policies:**

a. Finland's National Core Curriculum for Basic Education:
Finland's curriculum includes a strong emphasis on environmental education, integrating sustainability themes across various subjects. It aims to develop students' understanding of ecological systems, sustainable development, and responsible citizenship.
b. Germany's National Strategy for Education for Sustainable Development (ESD):
Germany's strategy focuses on integrating sustainability education into all levels of education. It emphasizes the development of environmental awareness, critical thinking, and action competence among students to address environmental challenges.
c. Australia's National Environmental Education Policy:
This policy promotes environmental education across all educational settings in Australia. It aims to develop students' environmental knowledge, skills, values, and behaviors to contribute to a sustainable future.

- **Common Elements in Policies:**

Several common elements can be observed in international and national policies promoting environmental sustainability in education:

- Integration: Policies emphasize the integration of sustainability themes into educational systems, encompassing curriculum, teaching practices, and institutional management.
- Multi-stakeholder Engagement: Policies recognize the importance of involving various stakeholders, including governments, educators, students, communities, and civil society organizations, in shaping and implementing sustainability education initiatives.
- Experiential Learning: Policies often emphasize hands-on, experiential learning approaches that connect students with the natural environment and foster a sense of environmental stewardship.
- Professional Development: Policies highlight the need for professional development opportunities for educators to enhance their capacity to deliver effective sustainability education.

These policies reflect a global recognition of the role of education in fostering environmental sustainability and equipping future generations with the knowledge and skills necessary for a sustainable future.

7.3.2 Case Studies of Successful Policy Implementation in Different Educational Settings

Here are a few case studies of successful policy implementation in different educational settings that have effectively promoted environmental sustainability:

- **Bhutan's Gross National Happiness (GNH) Policy:**

Bhutan has implemented the GNH policy, which emphasizes holistic well-being and sustainability. The education system in Bhutan incorporates the principles of GNH, with a focus on environmental conservation and sustainability education. Environmental issues are integrated into the curriculum, and students engage in hands-on learning experiences, such as tree planting and waste management projects. The policy has contributed to Bhutan's commitment to maintaining at least 60% forest coverage and has helped foster a sense of environmental stewardship among students.

- **New Zealand's Enviroschools Program:**

The Enviroschools program in New Zealand promotes sustainability education by engaging schools and their communities. The program supports schools in integrating sustainability into their curriculum, infrastructure, and culture. It encourages experiential learning, eco-friendly practices, and community partnerships. The program has been successful in fostering environmental awareness, empowering students to take action, and creating sustainable school environments [17].

- **Ontario's EcoSchools Program, Canada:**
Ontario's EcoSchools program integrates environmental education and sustainable practices into school operations and curriculum. It focuses on areas such as energy conservation, waste reduction, water conservation, and ecological literacy. The program provides schools with guidelines, resources, and recognition for their sustainability efforts. It has resulted in reduced energy consumption, increased waste diversion rates, and improved student engagement in environmental initiatives [18].

- **Costa Rica's Environmental Education Strategy:**
Costa Rica has implemented a comprehensive environmental education strategy that promotes sustainability and biodiversity conservation. The strategy integrates environmental education across all levels of the education system, including curriculum, teacher training, and community engagement. It has led to increased environmental awareness, community participation, and sustainable practices in schools and communities [19].

These case studies demonstrate the successful implementation of policies and programs that promote environmental sustainability in different educational settings. They showcase the importance of integrating sustainability into curriculum, engaging stakeholders, providing resources and support, and fostering experiential learning opportunities.

7.4 Curriculum Integration and Pedagogy

Integrating environmental education into curricula involves incorporating sustainability concepts and practices across subjects and grade levels. Here are some approaches and strategies for achieving this integration:

- Cross-curricular Connections: Identify opportunities to integrate environmental themes and concepts into various subjects, such as science, social studies, language arts, mathematics, and even arts and physical education. Design interdisciplinary projects and activities that allow students to explore environmental issues from different perspectives.

- Project-Based Learning: Engage students in project-based learning experiences that involve real-world environmental challenges. Students can investigate local environmental issues, propose solutions, and take action to address them. This approach fosters critical thinking, problem-solving, and collaboration skills while deepening students' understanding of sustainability.
- Place-Based Education: Connect classroom learning to the local environment and community. Incorporate field trips, outdoor experiences, and community partnerships into the curriculum to provide students with hands-on opportunities to engage with and learn about their local ecosystems and sustainability initiatives.
- Inquiry-Based Learning: Encourage students to ask questions, investigate, and explore environmental topics through inquiry-based learning approaches. This method promotes curiosity, research skills, and a deeper understanding of sustainability issues.

To effectively teach green topics, educators can employ innovative pedagogical methods and utilize resources specifically designed for environmental education. Some examples include:

- Experiential Learning: Engage students in hands-on, outdoor, and experiential learning experiences. Use nature walks, field trips, environmental simulations, and outdoor experiments to enhance students' connection to the natural world and deepen their understanding of green topics.
- Technology-Enhanced Learning: Utilize digital tools, online platforms, and interactive resources to explore environmental issues. Virtual field trips, online simulations, data analysis tools, and multimedia resources can enhance students' engagement and understanding of sustainability concepts.
- Storytelling and Case Studies: Use storytelling, real-life case studies, and narratives to make green topics relatable and engaging for students. Highlight success stories, challenges, and inspiring examples of individuals or communities making a positive impact on the environment.
- Green Infrastructure and Sustainable Practices: Incorporate sustainable practices within the school environment itself, such as recycling programs, energy conservation initiatives,

school gardens, and eco-friendly infrastructure. Use these practices as teaching tools to demonstrate practical applications of sustainability concepts.

To effectively implement green policies in the classroom, teachers need adequate training and professional development opportunities. Here are some key considerations:

- Environmental Education Training: Provide teachers with training on environmental education pedagogies, curriculum integration, and instructional strategies specific to teaching sustainability topics. This training can enhance teachers' knowledge, confidence, and skills in delivering effective environmental education.
- Collaboration and Networking: Encourage teachers to collaborate and share best practices in environmental education. Facilitate networking opportunities, workshops, and conferences where teachers can exchange ideas, resources, and experiences.
- Continued Professional Development: Offer ongoing professional development opportunities focused on environmental education. Provide access to relevant courses, webinars, online resources, and expert-led training sessions to support teachers' continuous learning and growth in this area.
- Resource and Material Support: Provide teachers with access to high-quality resources, curriculum materials, teaching guides, and lesson plans specifically designed for environmental education. This support can help teachers effectively integrate green topics into their teaching practice.

By integrating environmental education into curricula, utilizing innovative pedagogical methods, and providing teacher training and professional development, educators can effectively implement green policies in the classroom and foster a culture of sustainability among students.

7.5 Infrastructure and Facilities

Designing and retrofitting sustainable school buildings and facilities is an essential aspect of promoting environmental sustainability in the education sector. Here are key considerations and strategies for designing and retrofitting sustainable school buildings:

- **Energy Efficiency:**
  - ➢ Incorporate energy-efficient design principles, such as proper insulation, efficient HVAC systems, and high-performance windows, to reduce energy consumption.
  - ➢ Install energy-efficient lighting, including LED fixtures, motion sensors, and daylight harvesting systems, to optimize energy use.
  - ➢ Integrate renewable energy systems, such as solar panels or wind turbines, to generate clean energy onsite.

- **Water Conservation:**
  - ➢ Implement water-efficient fixtures, such as low-flow faucets, toilets, and urinals, to minimize water usage.
  - ➢ Design rainwater harvesting systems to collect and reuse rainwater for non-potable purposes like landscape irrigation or toilet flushing.

- **Indoor Environmental Quality:**
  - ➢ Prioritize natural ventilation and daylighting to improve indoor air quality and enhance the learning environment.
  - ➢ Select non-toxic and low-emitting materials for construction and interior finishes to promote healthy indoor air quality.
  - ➢ Design spaces that promote connection to nature, such as green walls, indoor plants, or outdoor learning areas.

- **Sustainable Materials and Waste Management:**
  - ➢ Use sustainable building materials with low embodied energy, such as recycled content materials, locally sourced materials, or sustainably harvested wood.
  - ➢ Incorporate waste management strategies, such as recycling stations and composting facilities, to minimize waste generation and promote recycling.

- **Accessibility and Inclusivity:**
  - ➢ Ensure that school buildings and facilities are designed to be accessible and inclusive for students with disabilities, including ramps, elevators, and accessible washrooms.

  - ➢ Create flexible spaces that can accommodate different learning needs and promote inclusive education.

- **Educational Opportunities:**
  - ➢ Use the school building and facilities as teaching tools to educate students about sustainable design principles and practices.
  - ➢ Incorporate sustainability features into the curriculum, allowing students to monitor energy consumption, analyze data, and propose improvements for resource conservation.

- **Long-term Monitoring and Maintenance:**
  - ➢ Implement systems for monitoring energy, water, and indoor air quality to identify areas for improvement and ensure ongoing sustainability performance.
  - ➢ Develop a maintenance plan to regularly inspect, maintain, and optimize the performance of sustainable building systems and features.

When retrofitting existing school buildings, consider conducting an energy audit to identify energy inefficiencies and prioritize retrofitting measures. Engage with architects, engineers, sustainability consultants, and stakeholders to develop a comprehensive sustainability plan aligned with the school's goals and budget. By incorporating sustainable design principles, schools can create healthier learning environments, reduce environmental impacts, and demonstrate sustainable practices to students and the community, fostering a culture of sustainability in education promotion of energy efficiency and renewable energy sources.

7.5.1 Waste Management and Recycling Initiatives in Educational Institutions

Waste management and recycling initiatives in educational institutions play a crucial role in promoting environmental sustainability and teaching students about responsible waste practices. These initiatives aim to minimize waste generation, promote recycling, and create a culture of sustainability within educational settings. Here are some common waste management and recycling initiatives implemented in educational institutions:

- Recycling Programs: Educational institutions often establish comprehensive recycling programs that include the collection and separation of recyclable materials such as paper, plastic, glass, and metal. Recycling bins are strategically placed throughout the campus, including in classrooms, hallways, cafeterias, and outdoor areas, to encourage students and staff to dispose of recyclable items correctly.

- Composting Initiatives: Some educational institutions implement composting initiatives to divert organic waste from landfill and turn it into nutrient-rich compost. Composting programs involve the collection of food scraps, yard waste, and other organic materials, which are then composted on-site or through partnerships with local composting facilities. The resulting compost can be used in school gardens or landscaping projects.
- Waste Audits: Waste audits are conducted to assess the composition and volume of waste generated within educational institutions. These audits help identify opportunities for waste reduction, recycling improvements, and behavior change. By analyzing the results of waste audits, schools can tailor their waste management strategies and target specific areas for improvement.
- Awareness and Education Campaigns: Educational institutions implement awareness and education campaigns to educate students, staff, and the wider community about waste management and recycling. These campaigns may include workshops, presentations, posters, and educational materials that highlight the importance of waste reduction, recycling, and responsible consumption. Awareness campaigns can also involve student-led initiatives, such as waste reduction competitions or awareness events.
- Sustainable Purchasing and Packaging: Educational institutions can adopt sustainable purchasing practices by considering the environmental impact of products and packaging materials. This includes selecting products with minimal packaging, purchasing recycled or environmentally friendly supplies, and encouraging the use of reusable items instead of single-use products.
- Collaboration and Partnerships: Schools often collaborate with local recycling centers, waste management companies, or environmental organizations to enhance their waste management and recycling initiatives. These partnerships can provide valuable resources, expertise, and support in

implementing effective waste management practices and promoting sustainability.

- Infrastructure and Facilities: Educational institutions can invest in appropriate waste management infrastructure and facilities to facilitate waste separation and recycling. This includes the provision of recycling bins, composting systems, signage, and clear instructions on waste sorting. Infrastructure improvements can help streamline waste management processes and increase recycling rates.
- Implementing this waste management and recycling initiatives in educational institutions not only contributes to environmental conservation but also fosters a sense of responsibility and environmental stewardship among students. By actively involving students and staff in waste reduction and recycling efforts, educational institutions play a vital role in shaping sustainable behaviors and creating a greener future.

7.5.2 Case Studies Showcasing Exemplary Green Infrastructure Projects in Education

Here are a few case studies showcasing exemplary green infrastructure projects in the education sector:

• The Bullitt Center—Seattle, Washington, USA:
The Bullitt Center is a six-story building that serves as an environmental research and education center. It incorporates numerous green infrastructure features, including rainwater harvesting, solar panels, geothermal heating and cooling systems, composting toilets, and natural ventilation. The building produces more energy than it consumes and uses rainwater for all its water needs. It serves as a living laboratory for students and visitors, demonstrating sustainable building practices and environmental stewardship [20].

• Singapore American School—Singapore:
The Singapore American School (SAS) implemented an innovative green infrastructure project by transforming its campus into a sustainable learning environment. The project included features such as green roofs, rain gardens, permeable pavements, and bioswales to manage stormwater runoff and reduce the heat island effect. These green infrastructure elements

provide hands-on learning opportunities for students to understand the importance of sustainable design and environmental conservation [21].

• **Hanover School District—Hanover, New Hampshire, USA:**
The Hanover School District in New Hampshire implemented sustainable infrastructure projects in its schools, including energy-efficient buildings, solar panels, and geothermal heating and cooling systems. The district's sustainability efforts resulted in significant energy cost savings and reduced environmental impact. The projects also served as educational tools, engaging students in learning about renewable energy and environmental stewardship [22].

7.6 Student Engagement and Participation

7.6.1 Student-Led Initiatives and Organizations Promoting Environmental Awareness and Action

Here are some examples of student-led initiatives and organizations that promote environmental awareness and action:

- Sunrise Movement: Sunrise Movement is a youth-led organization focused on advocating for political action on climate change. They organize campaigns, protests, and lobby efforts to promote renewable energy, sustainable policies, and climate justice.
- Roots & Shoots: Roots & Shoots is a global youth-led organization founded by Jane Goodall. It empowers young people to become environmental leaders through community-based projects that address issues like habitat conservation, waste reduction, and sustainable agriculture.
- Green Schools Initiative: The Green Schools Initiative engages students, teachers, and parents in promoting sustainability within schools. They provide resources, tools, and guidance to help schools reduce their environmental impact and educate students about sustainable practices.
- Youth Climate Strike: Youth Climate Strike is a movement that originated with the Fridays for Future protests started by Swedish activist Greta Thunberg. It involves students organizing strikes and walkouts to demand action on climate change and raise awareness about the urgency of the issue.

- Eco-Schools: Eco-Schools is an international program that encourages student-led environmental action in schools. It provides a framework for students to develop sustainability projects, reduce waste, conserve energy, and promote eco-friendly practices within their school communities.
- Students for Environmental Action (SEA): SEA is a student-led organization that promotes environmental awareness and sustainability initiatives on college campuses. They organize events, workshops, and campaigns to engage students in environmental issues and encourage sustainable behavior.
- Earth Guardians: Earth Guardians is a global organization led by young activists who work to address environmental and social justice issues. They empower young people to become leaders in their communities, advocating for sustainable solutions and positive change.

7.7 Collaboration and Partnerships

Collaboration among educational institutions, government agencies, NGOs, and the private sector is of paramount importance for addressing environmental sustainability challenges. Here are some reasons highlighting the significance of such collaboration:

- Diverse expertise and resources: Each stakeholder brings unique knowledge, skills, and resources to the table. Educational institutions offer expertise in teaching and learning, government agencies provide regulatory frameworks and policy expertise, NGOs contribute advocacy skills and community engagement, and the private sector brings technological innovations and financial resources. Collaborating allows for the pooling of these diverse resources and expertise to tackle complex environmental sustainability issues more effectively.
- Holistic and comprehensive solutions: Environmental sustainability challenges are multi-faceted and require holistic approaches. Collaboration among different stakeholders ensures a comprehensive understanding of the issues and the development of integrated solutions. By working together, educational institutions, government agencies, NGOs, and the private sector can identify synergies, leverage each other's

strengths, and address multiple dimensions of sustainability, such as social, economic, and environmental aspects.

- Leveraging networks and influence: Collaboration allows stakeholders to tap into each other's networks and spheres of influence. Educational institutions have access to students and young minds, government agencies have regulatory authority, NGOs have grassroots connections, and the private sector has market reach. By collaborating, these stakeholders can amplify their impact and reach a wider audience, facilitating the dissemination of knowledge, raising awareness, and mobilizing action for environmental sustainability.
- Policy development and implementation: Government agencies play a vital role in shaping policies and regulations related to environmental sustainability. Collaboration with educational institutions, NGOs, and the private sector ensures that policies are well-informed, taking into account scientific expertise, societal needs, and economic considerations. It also enhances the implementation of policies by leveraging the resources, knowledge, and community engagement capacities of other stakeholders.
- Scaling up initiatives: Collaborative efforts have the potential to scale up initiatives and make a larger impact. Educational institutions can incorporate sustainability education into curricula, NGOs can implement community-based projects, the private sector can invest in sustainable technologies, and government agencies can provide support and create an enabling environment. Collaboration enables the sharing of best practices, replication of successful initiatives, and scaling up of efforts to achieve broader environmental sustainability goals Examples of successful partnerships and networks for promoting green policies in education.

7.8 Monitoring, Evaluation, and Reporting

7.8.1 Importance of Monitoring and Evaluating the Implementation and Impact of Green Policies

Monitoring and evaluating the implementation and impact of green policies is crucial for several reasons:

- Accountability and transparency: Monitoring and evaluation provide a mechanism for holding policymakers and stakeholders accountable for their commitments and actions. It ensures transparency in the implementation of green policies and allows for public scrutiny. By regularly assessing progress, shortcomings can be identified and addressed, fostering a culture of accountability and enhancing public trust.
- Effectiveness assessment: Monitoring and evaluation help assess the effectiveness of green policies in achieving their intended outcomes. It allows policymakers to identify what works and what does not, enabling them to make informed decisions about policy adjustments, resource allocation, and interventions. By understanding the impact of policies, governments can refine and improve their strategies to maximize environmental benefits.
- Learning and knowledge sharing: Monitoring and evaluation generate valuable insights and lessons learned about the implementation of green policies. It provides an opportunity to identify successful practices, innovative approaches, and potential barriers or challenges. This knowledge can be shared across different levels of governance, sectors, and regions, fostering mutual learning and improving future policy design and implementation.
- Policy refinement and adaptation: Monitoring and evaluation provide feedback on the effectiveness of green policies, allowing policymakers to refine and adapt their strategies based on evidence and data. It helps identify areas where policy design or implementation can be improved, ensuring that policies remain relevant and responsive to changing environmental challenges and societal needs.
- Resource optimization: Effective monitoring and evaluation help optimize the allocation of resources. By assessing the impact of green policies, policymakers can identify areas of high impact and prioritize resource allocation accordingly. It ensures that resources are directed toward initiatives and interventions that deliver the most significant environmental benefits, maximizing the efficient use of limited resources.
- Stakeholder engagement and participation: Monitoring and evaluation processes can involve multiple stakeholders, including government agencies, NGOs, local communities, and affected individuals. By engaging these stakeholders in

the evaluation process, their perspectives, insights, and concerns can be taken into account. This participatory approach fosters ownership, collaboration, and collective responsibility, enhancing the legitimacy and acceptance of green policies.

7.8.2 Key Indicators and Evaluation Frameworks for Assessing Sustainability in Education

When assessing sustainability in education, there are several key indicators and evaluation frameworks that can be used. Here are some commonly utilized indicators and frameworks:

- Sustainable Development Goals (SDGs): The SDGs provide a comprehensive framework for assessing sustainability in various sectors, including education. Indicator 4.7 specifically focuses on education for sustainable development (ESD), emphasizing the integration of sustainability themes into the curriculum, teaching practices, and learning outcomes.
- Environmental Literacy Assessments: These assessments measure students' knowledge, understanding, and skills related to environmental concepts and issues. They assess competencies such as ecological knowledge, environmental decision-making, and behavior change. Examples include the National Environmental Education Foundation's Environmental Literacy Assessment and the Environmental Literacy Framework developed by the North American Association for Environmental Education.
- Carbon Footprint: This indicator measures the greenhouse gas emissions associated with educational institutions. It assesses emissions from energy consumption, transportation, waste management, and other activities. Evaluating and reducing the carbon footprint of educational institutions contributes to sustainability goals.
- Waste Management: Assessing waste management practices in educational institutions focuses on waste reduction, recycling, and proper disposal. Key indicators include waste diversion rates, recycling rates, and implementation of waste reduction strategies like composting and reusing materials.

- Biodiversity and Habitat Conservation: Evaluating efforts to conserve biodiversity and protect habitats can involve indicators such as the presence of native plant species, wildlife monitoring, creation of habitat areas on school grounds, and participation in conservation projects in partnership with local organizations.
- Sustainable Campus Operations: This indicator focuses on the sustainable management of educational institution campuses. It includes criteria such as energy efficiency, water conservation, sustainable transportation options, green building certifications (e.g., LEED), and sustainable procurement practices.
- Community Engagement: Assessing community engagement measures the level of involvement of educational institutions with local communities in sustainability initiatives. Indicators can include partnerships with community organizations, participation in community-based environmental projects, and outreach activities aimed at raising awareness and promoting sustainable practices beyond the campus.
- Social Justice and Equity: Evaluating sustainability in education also involves assessing the extent to which educational institutions promote social justice and equity. Indicators can include inclusive practices, diversity in curriculum and resources, and efforts to address environmental justice issues.

7.8.3 Reporting Mechanisms and Accountability in Relation to Green Policies

Reporting mechanisms and accountability play a crucial role in ensuring the effective implementation and monitoring of green policies. Here are some common mechanisms and strategies used for reporting and accountability:

- Reporting Guidelines and Standards: Governments, international organizations, and industry bodies often develop reporting guidelines and standards to facilitate the consistent reporting of sustainability-related information. Examples include the Global Reporting Initiative (GRI), the Sustainability Accounting Standards Board (SASB), and the Carbon Disclosure Project (CDP). These frameworks provide a structured approach to measuring, disclosing, and communicating environmental performance and sustainability efforts.

- Mandatory Reporting Requirements: Governments can enforce mandatory reporting requirements for organizations, requiring them to disclose their environmental performance, greenhouse gas emissions, and other sustainability-related information. These regulations establish a legal obligation for organizations to report and create a basis for accountability.
- Voluntary Reporting Initiatives: Voluntary reporting initiatives encourage organizations to disclose their environmental performance voluntarily. Examples include the United Nations Global Compact (UNGC) and the CDP. By participating in these initiatives, organizations signal their commitment to transparency and accountability, and their reports can be used to assess their sustainability efforts.
- Independent Verification and Assurance: Independent verification and assurance provide credibility and trust to sustainability reports. External auditors or specialized verification organizations assess the accuracy and completeness of reported data, ensuring compliance with reporting standards. This process enhances accountability and confidence in the reported information.
- Stakeholder Engagement and Feedback: Engaging stakeholders, including communities, civil society organizations, and shareholders, promotes accountability. Gathering input from stakeholders, conducting consultations, and responding to their concerns and feedback are important components of accountability mechanisms. Stakeholder engagement can help identify gaps, areas for improvement, and hold organizations or governments accountable for their sustainability commitments.
- Performance Targets and Benchmarks: Setting performance targets and benchmarks is a way to measure progress and hold organizations accountable. Green policies often include specific targets related to environmental performance, energy efficiency, waste reduction, or carbon emissions. Monitoring progress toward these targets enables stakeholders to assess the effectiveness of policies and identify areas where improvements are needed.

- Compliance and Enforcement: Governments can establish compliance mechanisms and enforcement measures to ensure adherence to green policies. This may involve penalties or incentives to drive compliance, inspections, and audits to monitor compliance, and legal mechanisms to address non-compliance.
- Transparency and Public Reporting: Transparency is a fundamental aspect of accountability. Making information publicly available, such as through online platforms or annual reports, allows stakeholders to access and assess the performance of organizations or governments against their sustainability commitments. Transparent reporting enhances accountability and enables public scrutiny.

7.9 Challenges and Future Directions

Implementing green policies in education can face several challenges and obstacles. Here are some key ones:

- Limited Resources: Lack of financial resources and budget constraints can hinder the implementation of green policies in education. Green initiatives often require investments in infrastructure, training, curriculum development, and sustainable technologies. Limited funding can restrict the ability of educational institutions to adopt sustainable practices and integrate environmental education into their curriculum.
- Resistance to Change: Resistance to change, both within educational institutions and among stakeholders, can be a significant challenge. There may be resistance from teachers, administrators, and staff who are accustomed to traditional practices and perceive sustainability initiatives as additional burdens or disruptions to their routines. Overcoming resistance requires effective communication, professional development, and creating a culture of sustainability within the educational community.
- Lack of Awareness and Knowledge: Many educators, policymakers, and stakeholders may have limited awareness and knowledge about the importance of sustainability in education. This lack of understanding can hinder the adoption of green policies and practices. Providing training, professional

development opportunities, and awareness campaigns can help address this challenge and build capacity for sustainable education.

- Institutional Barriers: Institutional structures, policies, and regulations may pose obstacles to implementing green policies. Bureaucratic processes, rigid curriculum frameworks, and standardized testing requirements can limit flexibility and innovation in integrating sustainability into education. Overcoming these barriers often requires policy changes, advocacy, and collaboration among different stakeholders.
- Curriculum Integration: Integrating sustainability into the curriculum can be challenging due to the need for interdisciplinary approaches and changes in instructional practices. Finding the balance between existing academic requirements and incorporating sustainability themes and concepts can be complex. It requires curriculum revision, teacher training, and the development of appropriate teaching materials and resources.
- Lack of Coordination and Collaboration: Collaboration among various stakeholders, including educational institutions, government agencies, NGOs, and the private sector, is crucial for effective implementation. However, the lack of coordination, communication gaps, and fragmented efforts can impede progress. Building effective partnerships, establishing coordination mechanisms, and fostering collaboration among stakeholders are essential for successful implementation.
- Policy Fragmentation and Inconsistency: Green policies in education may suffer from fragmentation and inconsistency across different levels of governance. Misalignment between national, regional, and local policies can create confusion and hinder coordinated action. Harmonizing policies, establishing clear guidelines, and ensuring effective communication and coordination among different levels of governance are necessary to address this challenge.
- Evaluation and Monitoring: The lack of robust evaluation and monitoring mechanisms can make it challenging to assess the impact of green policies in education. Without proper evaluation, it becomes difficult to measure progress, identify gaps, and refine strategies. Establishing evaluation frameworks, data collection systems, and regular monitoring

processes are essential for evidence-based decision-making and continuous improvement.

7.9.1 Exploration of Potential Solutions and Strategies to Address These Challenges

To address the challenges in implementing green policies in education, several solutions and strategies can be employed. Here are some potential approaches:

- Advocacy and Awareness Campaigns: Conducting advocacy and awareness campaigns to promote the importance of sustainability in education is crucial. These campaigns can target educators, policymakers, administrators, parents, and students to raise awareness about the benefits of green policies and their relevance to the future well-being of society and the planet.
- Capacity Building and Professional Development: Providing training and professional development opportunities for educators is essential for integrating sustainability into the curriculum effectively. Training programs can focus on pedagogical approaches, curriculum design, and the use of interactive teaching methods that foster critical thinking and problem-solving skills related to sustainability.
- Collaboration and Partnerships: Building strong collaborations and partnerships among educational institutions, government agencies, NGOs, and the private sector is crucial. These partnerships can facilitate resource-sharing, knowledge exchange, and joint initiatives. Collaborations can also provide access to expertise, funding opportunities, and best practices in implementing green policies.
- Policy Coherence and Integration: Ensuring policy coherence and integration at different levels of governance is vital. It involves aligning national, regional, and local policies to support the implementation of green policies in education. This can be achieved through the development of comprehensive sustainability frameworks and the establishment of coordination mechanisms among relevant stakeholders.
- Financial Support and Incentives: Providing financial support, grants, and incentives to educational institutions can

help overcome resource constraints and encourage the adoption of green policies. Governments and organizations can offer funding opportunities specifically dedicated to sustainability initiatives in education. Incentives can include recognition programs, awards, or certifications that acknowledge and promote sustainable practices in educational institutions.

- Curriculum Integration and Innovation: Actively integrating sustainability across different subjects and disciplines within the curriculum is vital. This integration can be facilitated through curriculum revision, the development of interdisciplinary learning materials, and the inclusion of real-world sustainability challenges in teaching and learning activities. Embracing innovative approaches, such as project-based learning and experiential education, can also enhance student engagement and understanding of sustainability concepts.
- Monitoring and Evaluation Systems: Establishing robust monitoring and evaluation systems is crucial for assessing the effectiveness of green policies in education. This includes developing indicators, collecting data, and conducting regular evaluations to measure progress and identify areas for improvement. Evaluation findings can inform evidence-based decision-making and guide policy adjustments.
- Student Empowerment and Engagement: Empowering students as agents of change is important for promoting sustainability in education. Encouraging student-led initiatives, providing platforms for student participation in decision-making processes, and integrating student perspectives in policy development can enhance their engagement and foster a sense of ownership and responsibility toward environmental sustainability.
- Continuous Learning and Knowledge Sharing: Encouraging continuous learning and knowledge sharing among educators, policymakers, and stakeholders is vital. This can be achieved through conferences, workshops, online platforms, and networks dedicated to sustainability in education. Sharing best practices, lessons learned, and success stories can inspire others and facilitate the replication and scaling up of effective initiatives.

7.9.2 Emerging Trends and Innovations in the Field of Green Policies in Education

The field of green policies in education is continually evolving, with emerging trends and innovations that aim to enhance sustainability efforts. Here are some notable trends and innovations in this field:

- Education for Sustainable Development (ESD): ESD is an approach that seeks to integrate sustainability principles and practices into all aspects of education. It emphasizes the development of knowledge, skills, values, and attitudes that empower individuals to contribute to a sustainable future. ESD promotes interdisciplinary learning, critical thinking, and problem-solving related to environmental, social, and economic sustainability.
- Green Campus Initiatives: Green campus initiatives focus on creating sustainable and eco-friendly educational institutions. These initiatives encompass various aspects, including sustainable infrastructure design, energy efficiency measures, waste reduction and recycling programs, water conservation, sustainable transportation options, and the promotion of green spaces on campuses. Green campuses serve as models for sustainability and provide hands-on learning opportunities for students.
- Technology and Digital Learning: Technology plays a significant role in promoting green policies in education. Digital learning platforms and tools enable access to educational resources on sustainability; facilitate remote and flexible learning, and support collaboration and knowledge sharing among educators and students. Virtual simulations and augmented reality can also enhance experiential learning in environmental science and sustainability.
- Outdoor and Experiential Learning: Outdoor and experiential learning programs provide students with firsthand experiences in nature and the environment. These programs engage students in activities such as field trips, nature exploration, outdoor laboratories, and service-learning projects. They foster a deep connection with nature, promote environmental stewardship, and enhance students' understanding of sustainability concepts.

- Green Curriculum and Pedagogy: Green curriculum and pedagogical approaches focus on integrating sustainability into the curriculum across disciplines. This includes the development of sustainability-focused courses, project-based learning, problem-solving activities, and inquiry-based teaching methods that foster critical thinking and engagement with real-world environmental challenges.
- Community Engagement and Partnerships: Collaboration and engagement with local communities, NGOs, businesses, and government agencies are increasingly recognized as essential components of green policies in education. Partnerships allow educational institutions to leverage external expertise, resources, and community-based initiatives. They also provide opportunities for students to engage in meaningful sustainability projects and service-learning activities that address community needs.
- Sustainable Procurement and Operations: Educational institutions are placing greater emphasis on sustainable procurement practices and environmentally friendly operations. This includes sourcing eco-friendly products and services, implementing sustainable waste management systems, adopting energy-efficient technologies, and promoting sustainable transportation options.
- Social Justice and Equity in Sustainability: There is growing recognition of the interconnectedness between sustainability and social justice. Educational institutions are integrating social justice perspectives into green policies, emphasizing the importance of addressing inequities and promoting inclusivity in sustainability efforts. This involves incorporating diverse voices, addressing environmental justice issues, and ensuring equal access to sustainability education and resources for all students.
- Research and Data-Driven Approaches: Increased emphasis is being placed on research and data-driven approaches to inform green policies in education. This includes conducting research on the effectiveness of sustainability interventions, evaluating the impact of green policies, and utilizing data to guide evidence-based decision-making and continuous improvement.

7.9.3 Summary of the Main Findings and Insights From the Research

Research on green policies in education has highlighted several important findings and insights. Here is a summary of the main ones:

- Environmental education benefits: Numerous studies have demonstrated the positive impact of environmental education on students. It enhances their knowledge and understanding of environmental issues, fosters a sense of responsibility, and promotes sustainable behaviors.
- Integration across the curriculum: Green policies emphasize integrating environmental themes and concepts across different subjects, rather than confining them to a specific environmental education class. This approach helps students develop a holistic understanding of environmental issues and their interconnectedness with various disciplines.
- Teacher training and professional development: Effective implementation of green policies requires well-trained and knowledgeable teachers. Research emphasizes the importance of providing adequate training and professional development opportunities to educators, enabling them to incorporate environmental education into their teaching practices effectively.
- Experiential and outdoor learning: Hands-on experiences and outdoor learning play a crucial role in green education. Studies have shown that engaging students in practical activities, such as field trips, gardening, and environmental projects, enhances their understanding and connection with nature while fostering a sense of stewardship.
- School infrastructure and resource management: Green policies also focus on improving the environmental sustainability of school infrastructure and resource management. Research highlights the significance of energy-efficient buildings, renewable energy sources, waste reduction and recycling programs, and sustainable transportation options in educational institutions.
- Community engagement and partnerships: Collaborative efforts involving schools, local communities, and relevant organizations yield better outcomes in green education. Partnerships facilitate resource-sharing, community

involvement, and opportunities for students to apply their knowledge and skills in real-world contexts.

- Evaluation and assessment: Evaluating the effectiveness of green policies and educational initiatives is crucial for their continuous improvement. Research emphasizes the need for robust evaluation frameworks and assessment methods to measure the impact of green education on students' knowledge, attitudes, and behaviors.
- Equity and inclusion: Green policies should address equity and inclusion to ensure that environmental education benefits all students, regardless of their socioeconomic status or background. Research suggests that integrating culturally responsive approaches, addressing environmental justice issues, and incorporating diverse perspectives enhance the effectiveness and relevance of green education.

Overall, research on green policies in education underscores the importance of integrating environmental education across the curriculum, providing professional development for teachers, emphasizing experiential learning, improving school infrastructure, fostering community partnerships, evaluating outcomes, and promoting equity and inclusion. By implementing these findings and insights, educational institutions can contribute to building a more environmentally sustainable and responsible society.

7.9.4 Contributions of the Study to the Field of Green Policies in Education

The study on green policies in education makes several valuable contributions to the field. Some of its key contributions include:

- Knowledge enhancement: The study consolidates existing research and knowledge on green policies in education, providing a comprehensive overview of the subject. By reviewing and synthesizing various studies, it helps researchers, policymakers, and educators deepen their understanding of the topic and identify areas for further exploration.
- Evidence-based insights: The study offers evidence-based insights into the effectiveness of green policies in education. By examining empirical research, it provides a solid

foundation for designing and implementing effective environmental education initiatives. Policymakers can use these insights to make informed decisions and allocate resources appropriately.

- Practical recommendations: The study provides practical recommendations for educators, schools, and policymakers to implement green policies effectively. These recommendations are based on the synthesis of research findings and best practices, offering actionable steps to integrate environmental education, improve infrastructure, engage communities, and promote equity and inclusion.
- Professional development guidance: The study emphasizes the importance of teacher training and professional development in green education. By highlighting the need for educators to acquire the necessary knowledge and skills, it guides the development of professional development programs that support teachers in integrating environmental concepts into their teaching practices.
- Evaluation frameworks: The study emphasizes the importance of evaluating the impact of green policies and educational initiatives. It contributes to the field by discussing evaluation frameworks and assessment methods that can be used to measure the effectiveness of environmental education programs. These frameworks help stakeholders assess outcomes, identify areas for improvement, and ensure accountability.
- Equity and inclusion considerations: The study highlights the significance of addressing equity and inclusion in green policies. By exploring the importance of cultural responsiveness, environmental justice, and diverse perspectives, it encourages policymakers and educators to develop inclusive approaches that consider the needs and experiences of all students, fostering a more equitable and diverse environmental education.
- Interdisciplinary perspective: The study promotes an interdisciplinary perspective by emphasizing the integration of environmental education across various subjects. By encouraging educators to incorporate environmental themes into different disciplines, it supports a comprehensive and interconnected understanding of environmental

issues, contributing to the development of well-rounded and environmentally conscious individuals.

By providing a synthesis of existing knowledge, evidence-based insights, practical recommendations, and a focus on equity and inclusion, the study makes important contributions to the field of green policies in education. It helps shape educational practices and policies, guiding stakeholders toward more effective and sustainable approaches to environmental education.

7.9.5 Implications for Policy, Practice, and Further Research

The study on green policies in education has several implications for policy, practice, and further research in the field. Here are some of the key implications:

Policy Implications:

- Integration and support: Policies should prioritize the integration of environmental education across the curriculum, providing guidelines and support for schools to incorporate environmental themes into various subjects. Policy frameworks should also allocate resources and funding to support the implementation of green policies and initiatives.
- Professional development: Policies should emphasize the importance of teacher training and professional development in environmental education. They should provide opportunities for educators to acquire the necessary knowledge and skills to effectively integrate environmental concepts into their teaching practices.
- Infrastructure and resource management: Policies should promote sustainable practices in school infrastructure and resource management. This includes encouraging energy-efficient buildings, renewable energy sources, waste reduction and recycling programs, and sustainable transportation options.
- Community engagement: Policies should foster partnerships and collaboration between schools, local communities, and relevant organizations. They should encourage community involvement, resource-sharing, and opportunities for students to apply their knowledge and skills in real-world contexts.

Practice Implications:

- Curriculum integration: Educators should integrate environmental themes across various subjects, connecting them to real-world issues and challenges. They should develop interdisciplinary projects and activities that promote a holistic understanding of environmental issues.
- Experiential learning: Educators should provide hands-on experiences and outdoor learning opportunities to enhance students' understanding and connection with nature. They should engage students in practical activities, such as field trips, gardening, and environmental projects.
- Culturally responsive approaches: Educators should adopt culturally responsive approaches that consider the diverse backgrounds and experiences of students. They should incorporate diverse perspectives, address environmental justice issues, and promote inclusivity in environmental education.
- Evaluation and assessment: Educators should develop and implement robust evaluation frameworks and assessment methods to measure the impact of green policies and educational initiatives. This will help monitor progress, identify areas for improvement, and ensure the effectiveness of environmental education programs.

7.9.6 Further Research Implications

- Long-term impact: Further research is needed to assess the long-term impact of green policies and environmental education on students' knowledge, attitudes, and behaviors. Longitudinal studies can provide insights into the lasting effects of environmental education on individuals and communities.
- Equity and inclusion: More research is needed to explore the effectiveness of strategies and approaches to promote equity and inclusion in environmental education. This includes investigating the impact of culturally responsive practices, addressing environmental justice issues, and ensuring that environmental education reaches marginalized and underserved communities.

- Teacher preparation and support: Research can focus on identifying effective approaches for teacher training and professional development in environmental education. This includes investigating the most impactful methods for building educators' knowledge, skills, and confidence in integrating environmental concepts into their teaching.
- Global perspectives: Further research can explore the cultural and contextual factors influencing the implementation and effectiveness of green policies in different regions and countries. Comparative studies can provide insights into best practices and lessons learned from diverse educational contexts.

By addressing these implications in policy, practice, and further research, the field of green policies in education can continue to evolve, leading to more effective and sustainable environmental education initiatives.

7.10 Conclusion

The research study on green policies in education draws several key conclusions based on its findings and analysis. These conclusions serve to summarize the main insights and implications of the study. Here is an example of how a research study conclusion on this topic could be formulated: In conclusion, the study on green policies in education highlights the significant role that environmental education plays in fostering sustainable behaviors, developing critical thinking skills, and promoting a sense of environmental stewardship among students. The integration of environmental concepts across the curriculum, coupled with experiential and outdoor learning, emerges as a powerful approach to enhance students' understanding of environmental issues and their connection with nature. The study underscores the importance of teacher training and professional development to equip educators with the necessary knowledge and skills to effectively integrate environmental education into their teaching practices. Furthermore, the study emphasizes the need for robust evaluation frameworks to assess the impact of green policies and educational initiatives, ensuring continuous improvement and accountability. The findings also emphasize the significance of community engagement, infrastructure improvements, and addressing equity and inclusion in environmental education. Overall, the study contributes to the growing body of knowledge in the field of green policies in education, providing evidence-based insights and practical recommendations for policymakers, educators, and researchers to enhance environmental education practices and policies, ultimately contributing to the creation of a more sustainable and environmentally conscious society.

References

1. UNESCO, *Education for Sustainable Development Goals: Learning Objectives*, UNESCO, Paris, 2017.
2. Mezirow, J., *Learning as Transformation: Critical Perspectives on a Theory in Progress*, Jossey-Bass, San Francisco, 2000.
3. UNESCO, *Final Report: Intergovernmental Conference on Environmental Education*, UNESCO, Tbilisi, 1978.
4. Sobel, D., *Place-Based Education: Connecting Classrooms and Communities*, The Orion Society, Great Barrington, 2004.
5. Meadows, D.H., *Thinking in Systems: A Primer*, Chelsea Green Publishing, White River Junction, 2008.
6. Dobson, A., *Citizenship and the Environment*, Oxford University Press, Oxford, 2003.
7. Sabatier, P.A. and Jenkins-Smith, H.C., *Policy Change and Learning: An Advocacy Coalition Approach*, Westview Press, Boulder, 1993.
8. DiMaggio, P.J. and Powell, W.W., The Iron Cage Revisited: Institutional Isomorphism and Collective Rationality in Organizational Fields. *Am. Sociological Rev.*, 48, 2, 147–160, 1983.
9. Rogers, E.M., *Diffusion of Innovations*, 5th, Free Press, New York, 2003.
10. Rhodes, R.A.W., Understanding Governance: Policy Networks, Governance, Reflexivity, and Accountability. *Public Adm.*, 75, 3, 493–515), 1997.
11. Ball, S.J., *Global Education Inc.: New Policy Networks and the Neoliberal Imaginary*, Routledge, New York, 2012.
12. Lasswell, H.D., The Policy Orientation, in: *The Policy Sciences: Recent Developments in Scope and Method*, D. Lerner, and H.D. Lasswell, (Eds.), pp. 3–15, Stanford University Press, Palo Alto, 1951.
13. Day, C. and Gu, Q., *The New Lives of Teachers*, Routledge, London, 2010.
14. Howlett, M. and Ramesh, M., *Studying Public Policy: Policy Cycles and Policy Subsystems*, Oxford University Press, Oxford, 2003.
15. Van Meter, D.S. and Van Horn, C.E., The Policy Implementation Process: A Conceptual Framework. *Adm. Soc.*, 6, 4, 445–488, 1975.
16. Patton, M.Q., *Developmental Evaluation: Applying Complexity Concepts to Enhance Innovation and Use*, 2011. Ura, K. and Galay, K., Gross National Happiness and Education in Bhutan. *J. Bhutan Stud.*, 11, 1–30, 2004.

17. McGovern, M. and Duncan, J., The New Zealand Enviroschools Programme: Resilience and Sustainability in Education, in: *Sustainability Science: An Introduction*, Gray, D. and Colucci-Gray, M. (Eds.), pp. 127–143, 2014.
18. Imai, J., The Ontario EcoSchools Program: A Case Study in School-Based Sustainability Education and Professional Learning. *J. Environ. Educ.*, 46, 4, 202–220, 2015.
19. UNESCO, *Costa Rica's Environmental Education Strategy: Learning to Care for Our Common Future*, UNESCO, Paris, 2014.
20. Moffitt, B. and Hill, B., Living Buildings: The Bullitt Center, in: *Green Infrastructure Implementation*, pp. 3–13, Springer, Cham, 2016.
21. Law, D.C. and Lee, M.C., Developing a Campus-Based Green Infrastructure Planning Process for K-12 Educational Facilities: A Case Study of Singapore American School, in: *Green Infrastructure Implementation*, pp. 165–185, Springer, Cham, 2013.
22. EPA, Achieving 100% Renewable Energy in Hanover, Environmental Protection Agency, Washington, D.C., 2017.

8

Green Engineering in Automobile Sector

Harjit Singh*, Mukhtiar Singh, Jasvinder Singh and Maninder Singh

School of Mechanical Engineering, Lovely Professional University, Punjab, India

Abstract

This book chapter on green engineering in the automobile sector highlights the need for sustainable practices in the automotive industry. With increasing concerns over climate change, pollution, and energy consumption, green engineering has become a critical aspect of automobile design, manufacturing, and operation. The chapter provides an overview of green engineering practices in the automobile sector, covering design considerations, sustainable manufacturing practices, energy-efficient operation, and end-of-life options for automobiles. Case studies are presented to showcase successful examples of green engineering practices in the industry. The chapter concludes with a discussion on the potential benefits of continued implementation of green engineering in the automobile sector, and future directions for the industry to pursue. Overall, this book chapter emphasizes the importance of sustainable engineering practices in the automobile industry to mitigate environmental impact and improve the sustainability of transportation.

Keywords: Fuel consumption, lightweight materials, environmental impact, product life cycle, greenhouse gas emissions, lean manufacturing

8.1 Introduction

The automobile industry has a significant impact on the environment, contributing to air pollution, greenhouse gas emissions, and natural resource depletion. In response, there has been a growing emphasis on sustainable practices in the automotive sector, including the implementation of green engineering principles.

**Corresponding author*: Harjit.17005@lpu.com.in

Chandan Deep Singh, Talwinder Singh and Davinder Singh (eds.) Digitization and Manufacturing Performance: An Environmental Perspective, (211–234) © 2025 Scrivener Publishing LLC

Green engineering is defined as "the design, commercialization, and use of processes and products that are feasible and economical while minimizing the generation of pollution at the source and the risk to human health and the environment" [1]. In the context of the automobile sector, green engineering involves the development of sustainable practices throughout the entire product life cycle, from design to end-of-life disposal.

Green engineering practices in the automobile sector have become increasingly important due to the environmental challenges facing the industry. The transportation sector is responsible for around one-quarter of global greenhouse gas emissions, and road vehicles contribute to a significant portion of that [2]. In addition, the depletion of non-renewable resources and the need to reduce waste and pollution have led to an increased demand for sustainable engineering practices in the automobile sector.

In response, automobile manufacturers are implementing green engineering practices throughout their supply chains to reduce the environmental impact of their products. This includes the use of alternative energy sources, such as electric and hybrid vehicles, the development of lightweight materials to reduce fuel consumption, and the use of sustainable manufacturing processes to reduce waste and emissions.

This book chapter aims to provide an overview of green engineering practices in the automobile sector, covering design, manufacturing, and operation. Case studies are presented to showcase successful examples of green engineering practices in the industry. The chapter also highlights the potential benefits of continued implementation of green engineering in the automobile sector and discusses future directions for the industry to pursue.

- **Background on the need for green engineering in the automobile sector**

The automobile industry is a significant contributor to environmental degradation and climate change. According to the International Energy Agency (IEA), the transportation sector is responsible for approximately 24% of global energy-related greenhouse gas emissions, and road vehicles account for the majority of those emissions [2]. In addition to greenhouse gas emissions, automobiles also contribute to air pollution, natural resource depletion, and waste generation.

To address these environmental challenges, there has been a growing emphasis on sustainable practices in the automobile sector. Green engineering, which aims to minimize the environmental impact of products and processes through the use of sustainable design, manufacturing, and operation, has become an important aspect of sustainability in the automotive industry.

The need for green engineering in the automobile sector has become increasingly urgent as global environmental challenges continue to escalate. The depletion of non-renewable resources and the need to reduce waste and pollution have led to an increased demand for sustainable engineering practices in the automobile sector. The automotive industry is also facing increased regulation and pressure from consumers to reduce their environmental impact.

In response, automobile manufacturers are implementing green engineering practices throughout their supply chains to reduce the environmental impact of their products. This includes the use of alternative energy sources, such as electric and hybrid vehicles, the development of lightweight materials to reduce fuel consumption, and the use of sustainable manufacturing processes to reduce waste and emissions.

The need for green engineering in the automobile sector is driven by the need to reduce the environmental impact of transportation and to improve the sustainability of the automotive industry. By implementing sustainable practices throughout the product life cycle, automobile manufacturers can reduce their environmental footprint and meet the demands of consumers and regulators for more sustainable products.

- **Definition of green engineering in the context of the automobile sector**

Green engineering is a design and engineering philosophy that aims to minimize the environmental impact of products and processes by integrating sustainability into every stage of a product's life cycle. In the context of the automobile sector, green engineering involves developing sustainable practices that reduce the environmental impact of vehicles, from design and manufacturing to end-of-life disposal.

According to the US Environmental Protection Agency (EPA), green engineering "is the practice of designing, developing, and implementing products, processes, and systems that minimize the environmental impact throughout the entire life cycle of the product or process" [3]. Green engineering aims to reduce the use of non-renewable resources, minimize waste generation, and reduce emissions and pollutants throughout the life cycle of a product.

In the automobile sector, green engineering includes a range of practices, including the use of alternative energy sources, such as electric and hybrid vehicles, the development of lightweight materials to reduce fuel consumption, and the use of sustainable manufacturing processes to reduce waste and emissions [6]. Green engineering in the automobile sector also involves the development of end-of-life options for vehicles, such as recycling and reuse, to minimize waste and maximize the use of materials.

Green engineering in the automobile sector is an important aspect of sustainability, as it enables the development of more environmentally friendly vehicles and reduces the environmental impact of the automotive industry.

- **Importance of green engineering in the automobile sector**

Green engineering in the automobile sector is becoming increasingly important as the need for sustainable practices grows more urgent. The automobile industry is a significant contributor to environmental degradation and climate change, and the use of traditional manufacturing practices and fossil fuels in vehicles has negative impacts on the environment and human health.

By integrating green engineering practices into the design, manufacturing, and end-of-life disposal of vehicles, automobile manufacturers can reduce their environmental footprint and create more sustainable products. The benefits of green engineering in the automobile sector include:

- ➢ Reduced greenhouse gas emissions: The transportation sector is responsible for a significant portion of global greenhouse gas emissions, and the majority of those emissions come from road vehicles. By developing and producing more sustainable vehicles, the automobile industry can help reduce the carbon footprint of transportation and contribute to global efforts to mitigate climate change.
- ➢ Resource conservation: Green engineering in the automobile sector can help reduce the consumption of non-renewable resources, such as fossil fuels and metals, by developing more efficient and sustainable manufacturing processes, as well as creating vehicles that use alternative energy sources.
- ➢ Improved air quality: Vehicles are a major source of air pollution, which has negative impacts on human health and the environment. Green engineering practices can reduce the emissions of pollutants such as nitrogen oxides, particulate matter, and volatile organic compounds, leading to improved air quality.
- ➢ Reduced waste: The automobile industry generates significant amounts of waste, including scrap metal, plastic, and hazardous materials. Green engineering practices can help reduce waste generation and promote the reuse and recycling of materials, reducing the environmental impact of the industry.

Green engineering in the automobile sector is important for promoting sustainability and reducing the environmental impact of transportation. By implementing sustainable practices throughout the product life cycle, automobile manufacturers can create more environmentally friendly products and contribute to a more sustainable future.

8.2 Green Engineering in Automobile Design

Green engineering in automobile design involves developing vehicles that are more sustainable and environmentally friendly. This includes reducing the environmental impact of the manufacturing process, using alternative materials and energy sources, and improving the efficiency of the vehicle's operation.

One approach to green engineering in automobile design is to focus on lightweight materials. Lightweight materials, such as carbon fiber, aluminum, and high-strength steel, can reduce the weight of the vehicle, improving fuel efficiency and reducing emissions. In addition, using recycled or sustainable materials in vehicle construction can reduce the environmental impact of the manufacturing process.

Another approach to green engineering in automobile design is to focus on the development of electric and hybrid vehicles. Electric vehicles produce no tailpipe emissions and have lower lifecycle emissions compared to traditional vehicles [5]. Hybrid vehicles combine an internal combustion engine with an electric motor, reducing fuel consumption and emissions.

Green engineering in automobile design also involves developing more efficient engines and powertrains, as well as using advanced technologies to improve vehicle aerodynamics. These approaches can help improve fuel efficiency and reduce emissions, while maintaining or improving vehicle performance.

Green engineering in automobile design is an important aspect of sustainability, as it enables the development of more environmentally friendly vehicles and reduces the environmental impact of the automotive industry.

- **Design considerations for green engineering**

Green engineering in the automobile sector requires careful consideration of design choices in order to achieve sustainability goals. Here are some design considerations for green engineering in the automobile sector:

- ➢ Material selection: Choosing sustainable and environmentally friendly materials is an important consideration in green engineering. This can include using recycled materials, reducing the use of toxic materials, and selecting materials with a lower environmental impact. For example, replacing traditional steel with lightweight materials like aluminum and carbon fiber can reduce the weight of the vehicle, improve fuel efficiency, and reduce emissions [6].

- ➢ Energy efficiency: Designing vehicles with energy efficiency in mind is another important consideration for green engineering. This can include using technologies like regenerative braking, which converts kinetic energy into electricity to charge the vehicle's battery. Using more efficient engines and powertrains can also improve fuel efficiency and reduce emissions [5].
- ➢ End-of-life considerations: Designing vehicles with end-of-life considerations in mind is important for reducing waste and promoting recycling. For example, designing vehicles with easily disassembled parts can facilitate the recycling of materials and reduce waste [6].
- ➢ Manufacturing processes: Green engineering in the automobile sector also involves designing manufacturing processes that are more sustainable and environmentally friendly. This can include using renewable energy sources like solar and wind power to power manufacturing facilities, as well as reducing waste and improving resource efficiency [4].

By considering these design considerations, green engineering in the automobile sector can promote sustainability and reduce the environmental impact of the industry.

- **Green materials selection for automobiles**

Green materials selection is an important consideration in the development of environmentally friendly automobiles. Here are some examples of green materials used in the automobile sector:

- ➢ Bioplastics: Bioplastics are a type of plastic made from renewable biomass sources, such as corn starch, sugarcane, and vegetable fats and oils. Bioplastics are biodegradable and have a lower environmental impact than traditional petroleum-based plastics [7].
- ➢ Recycled materials: Using recycled materials in vehicle manufacturing can reduce the environmental impact of the

industry. For example, recycled steel and aluminum can be used in vehicle construction, reducing the need for new materials [6].

- Natural fibers: Natural fibers, such as jute, hemp, and flax, can be used as alternatives to traditional synthetic fibers. These materials have a lower environmental impact and are biodegradable [7].
- Bio-based foams: Bio-based foams are a type of foam made from renewable materials, such as soybean oil and castor oil. These materials have a lower environmental impact than traditional petroleum-based foams [6].
- Bio-based adhesives: Bio-based adhesives are a type of adhesive made from renewable materials, such as soybean oil and corn starch. These adhesives have a lower environmental impact than traditional petroleum-based adhesives [6].

By selecting green materials for automobile manufacturing, the industry can reduce its environmental impact and promote sustainability.

- **Energy-efficient automobile design**

Energy-efficient automobile design is an important consideration in the development of environmentally friendly automobiles. Here are some examples of energy-efficient design strategies used in the automobile sector:

- Lightweight materials: Lightweight materials, such as aluminum, carbon fiber, and composites, can be used in vehicle design to reduce weight and improve fuel efficiency [6].
- Aerodynamics: Designing vehicles with good aerodynamics can improve fuel efficiency by reducing drag. This can include using features such as air dams, underbody covers, and streamlined body shapes [5].
- Powertrain efficiency: Improving powertrain efficiency is another important strategy for energy-efficient automobile design. This can include using technologies such as hybrid powertrains, regenerative braking, and engine start-stop systems [5].
- Electric and hybrid vehicles: Electric and hybrid vehicles are becoming increasingly popular due to their energy efficiency and low emissions. These vehicles use electric motors and batteries to reduce reliance on fossil fuels [5].

By incorporating these energy-efficient design strategies, the automobile industry can reduce its environmental impact and promote sustainability.

8.2.1 Green Engineering in Automobile Manufacturing

Green engineering in automobile manufacturing is an important aspect of promoting sustainability in the industry. Here are some examples of green engineering practices used in the automobile sector:

- Lean manufacturing: Lean manufacturing is a philosophy that focuses on minimizing waste and maximizing efficiency in the manufacturing process. This can include using methods such as just-in-time inventory and reducing energy usage in the manufacturing process [8].
- Renewable energy: The use of renewable energy sources, such as solar and wind power, can reduce the environmental impact of automobile manufacturing. Some automobile manufacturers have invested in renewable energy projects to power their manufacturing facilities [9].
- Water conservation: Water is an important resource in automobile manufacturing, and its conservation is an important consideration for green engineering. This can include using water-saving technologies and recycling water in the manufacturing process [8].
- Sustainable supply chain: The sustainability of the automobile manufacturing process depends not only on the manufacturing facility itself but also on the sustainability of the supply chain. This can include using green materials and promoting sustainable practices among suppliers [9].

By implementing these green engineering practices, the automobile industry can reduce its environmental impact and promote sustainability.

- **Sustainable manufacturing practices in the automobile industry**

Sustainable manufacturing practices are becoming increasingly important in the automobile industry as companies aim to reduce their environmental impact and promote sustainability. Here are some examples of sustainable manufacturing practices used in the automobile sector:

- Closed-loop manufacturing: Closed-loop manufacturing involves using materials that can be recycled or reused in the manufacturing process. This can reduce waste and promote sustainability [13].

- Zero-waste manufacturing: Zero-waste manufacturing involves eliminating waste from the manufacturing process entirely. This can include using materials that are biodegradable or compostable and recycling or repurposing waste products [12].
- Life cycle assessment: Life cycle assessment is a tool used to evaluate the environmental impact of a product over its entire life cycle, from production to disposal. This can help identify areas where improvements can be made to reduce the environmental impact of the product [10].
- Sustainable energy use: The use of renewable energy sources, such as solar and wind power, can reduce the environmental impact of manufacturing processes [11].

By incorporating these sustainable manufacturing practices, the automobile industry can reduce its environmental impact and promote sustainability.

- **Reducing waste and emissions in automobile manufacturing**

Reducing waste and emissions in automobile manufacturing is an important aspect of green engineering. Here are some examples of practices that can be used to reduce waste and emissions in the automobile manufacturing process:

- Lean manufacturing: Lean manufacturing is a philosophy that focuses on minimizing waste and maximizing efficiency in the manufacturing process. This can include reducing the amount of raw materials used, using just-in-time inventory systems, and reducing energy usage in the manufacturing process [8].
- Recycling and reuse: Recycling and reuse of materials can reduce waste in the manufacturing process. This can include recycling scrap metal, reusing water, and using recycled plastics in the manufacturing process [13].
- Low-emission manufacturing processes: The use of low-emission manufacturing processes, such as using water-based paints instead of solvent-based paints, can reduce emissions in the manufacturing process [8].

- Sustainable transportation: The transportation of materials and products in the automobile manufacturing process can contribute to emissions. Using sustainable transportation methods, such as electric or hybrid trucks, can reduce emissions [13].

By implementing these practices, the automobile industry can reduce waste and emissions in the manufacturing process and promote sustainability.

• The role of lean manufacturing in green engineering

Lean manufacturing is a production philosophy that focuses on reducing waste, improving efficiency, and optimizing resources. The principles of lean manufacturing align with the goals of green engineering, as both aim to reduce environmental impact and promote sustainability. Here are some examples of the role of lean manufacturing in green engineering:

- Waste reduction: Lean manufacturing techniques, such as just-in-time inventory and value stream mapping, can help identify and eliminate waste in the manufacturing process, resulting in reduced resource usage and environmental impact [15].
- Energy efficiency: Lean manufacturing principles can also lead to increased energy efficiency by optimizing processes and reducing unnecessary energy consumption [14].
- Continuous improvement: Lean manufacturing promotes a culture of continuous improvement, which can lead to the development and implementation of more sustainable practices over time [15].
- Sustainability-focused design: By integrating lean manufacturing principles into the design process, products can be designed with sustainability in mind, resulting in reduced environmental impact over the entire product life cycle [16].

Incorporating lean manufacturing principles into green engineering can help reduce environmental impact and promote sustainability in the manufacturing process.

8.2.2 Green Engineering in Automobile Operations

Green engineering in automobile operations refers to the use of sustainable practices in the maintenance and operation of vehicles. Here are some examples of green engineering practices that can be implemented in automobile operations:

- ➢ Eco-driving: Eco-driving techniques, such as smooth acceleration and braking, reducing idling time, and maintaining proper tire pressure, can help improve fuel efficiency and reduce emissions [18].
- ➢ Alternative fuels: Alternative fuels, such as electric, hybrid, or biofuels, can reduce emissions and dependence on fossil fuels [17].
- ➢ Maintenance and repair: Regular maintenance and repair can help keep vehicles operating efficiently and reduce emissions. Proper maintenance can include regular oil changes, air filter replacements, and tune-ups [18].
- ➢ Car-sharing and ride-sharing: Car-sharing and ride-sharing programs can reduce the number of vehicles on the road, which can help reduce emissions and traffic congestion [17].

Implementing these practices in automobile operations can help reduce environmental impact and promote sustainability.

- **Improving automobile efficiency**

Improving automobile efficiency is a key aspect of green engineering in the automobile sector. Here are some examples of techniques that can be used to improve automobile efficiency:

- ➢ Lightweight materials: Using lightweight materials, such as aluminum, carbon fiber, and composites, can reduce vehicle weight and improve fuel efficiency [20].
- ➢ Aerodynamics: Designing vehicles with aerodynamic features, such as rounded edges and streamlined shapes can reduce air resistance and improve fuel efficiency [19].
- ➢ Engine efficiency: Improving engine efficiency, such as through the use of hybrid or electric powertrains, can reduce fuel consumption and emissions [21].

- ➢ Intelligent transportation systems: Using intelligent transportation systems, such as adaptive cruise control and traffic signal coordination, can reduce traffic congestion and improve fuel efficiency [22].

Implementing these techniques can improve automobile efficiency and reduce environmental impact.

- **The role of electric vehicles in green engineering**

Electric vehicles (EVs) play a significant role in green engineering in the automobile sector. Here are some examples of the role of electric vehicles in green engineering:

- ➢ Reduction of greenhouse gas emissions: EVs produce zero tailpipe emissions, which can help reduce greenhouse gas emissions and improve air quality [24].
- ➢ Renewable energy integration: EVs can be charged using renewable energy sources, such as solar and wind power, which can reduce reliance on fossil fuels and promote sustainability [23].
- ➢ Energy efficiency: EVs are more energy efficient than traditional internal combustion engine vehicles, which can reduce energy consumption and costs [24].
- ➢ Technology innovation: The development and implementation of EV technology promotes innovation in the automotive industry and can lead to further advancements in green engineering [25].

Overall, the use of EVs can significantly contribute to reducing environmental impact and promoting sustainability in the automobile sector.

- **Reducing greenhouse gas emissions in the automobile sector**

Reducing greenhouse gas (GHG) emissions is a critical aspect of green engineering in the automobile sector. Here are some examples of strategies that can be used to reduce GHG emissions:

- ➢ Alternative fuels: The use of alternative fuels, such as biofuels, hydrogen, and natural gas, can help reduce GHG emissions from vehicles [26].

- Lightweight materials: Lightweight materials, such as aluminum and carbon fiber composites, can help reduce the weight of vehicles and improve fuel efficiency, which can in turn reduce GHG emissions [27].
- Engine efficiency improvements: Improving the efficiency of internal combustion engines through strategies such as turbocharging and downsizing can reduce GHG emissions [26].
- Electric vehicles: As mentioned earlier, electric vehicles produce zero tailpipe emissions and can significantly reduce GHG emissions [24].

Overall, a combination of these strategies can help reduce GHG emissions and promote sustainability in the automobile sector.

8.2.3 Green Engineering in Automobile End-of-Life

Green engineering also involves end-of-life considerations for automobiles, which includes strategies for reducing waste and promoting recycling. Here are some examples of strategies for green engineering in automobile end-of-life:

- Design for disassembly: Designing vehicles with disassembly in mind can make it easier to separate and recycle components at the end of their life [28].
- Recycling: The use of recycled materials in the manufacture of new vehicles can help reduce waste and the environmental impact of production [28].
- End-of-life management: Proper management of end-of-life vehicles, including disposal, recycling, and recovery of materials, is crucial for reducing waste and promoting sustainability [20].
- Circular economy principles: Adopting circular economy principles, such as closed-loop material flows and extended product lifetimes, can also help reduce waste and promote sustainability in the automobile sector [28].

Green engineering in automobile end-of-life involves a range of strategies for reducing waste and promoting recycling, and it is a crucial aspect of sustainable automobile design and production [29].

- **Sustainable disposal and recycling of automobiles**

The sustainable disposal and recycling of automobiles is an important aspect of green engineering. Here are some examples of strategies for sustainable disposal and recycling of automobiles:

  - De-pollution: Prior to recycling or disposal, end-of-life vehicles require de-pollution, which involves removing hazardous materials such as batteries, fluids, and refrigerants [30].
  - Material recovery: Automobile recycling involves recovering materials such as metals, plastics, and glass for use in the manufacture of new products [30].
  - Extended producer responsibility: Implementing extended producer responsibility programs can help ensure that automobile manufacturers are responsible for the end-of-life disposal and recycling of their products [31].
  - Certified recycling facilities: The use of certified recycling facilities can help ensure that end-of-life vehicles are recycled in an environmentally sound manner [30].

Overall, sustainable disposal and recycling of automobiles is crucial for reducing waste and promoting a circular economy in the automobile sector.

- **End-of-life options for electric vehicle batteries**

As electric vehicles (EVs) become more common, the end-of-life management of EV batteries is becoming an increasingly important aspect of green engineering. Here are some options for end-of-life management of EV batteries:

  - Second-life applications: EV batteries that are no longer suitable for use in vehicles can still have utility in second-life applications, such as stationary energy storage [32].
  - Recycling: Recycling is an important strategy for reducing waste and recovering valuable materials from EV batteries [32].
  - Repurposing: EV batteries can be repurposed for non-vehicle applications, such as providing power for off-grid homes or as backup power for critical infrastructure [32].
  - Landfill disposal: Landfill disposal should be a last resort for EV batteries, as it represents a missed opportunity for resource recovery [32].

The end-of-life management of EV batteries requires careful consideration to ensure that valuable materials are recovered and waste is minimized.

- **Extended producer responsibility for automobile manufacturers**

Extended Producer Responsibility (EPR) is a policy approach that aims to make producers responsible for the disposal and recycling of their products, including automobiles. Here are some references and citations related to EPR for automobile manufacturers:

- ➢ Extended Producer Responsibility [31]: A Guidance Manual for Governments. Retrieved from https://www.unenvironment.org/resources/report/extended-producer-responsibility-guidance-manual-governments.

  This manual by the United Nations Environment Programme (UNEP) provides guidance for governments on the implementation of EPR policies. It discusses the benefits of EPR, the key elements of EPR programs, and best practices for implementation.
- ➢ European Commission [43]. Extended Producer Responsibility. Retrieved from https://ec.europa.eu/environment/waste/producer_responsibility/index_en.htm.

  The European Commission provides information on EPR policies for various product categories, including automobiles. It outlines the key principles of EPR, the benefits of EPR, and the EU's legal framework for EPR.
- ➢ Extended producer responsibility in China [36]: Progress, challenges and policy implications. Journal of Cleaner Production, 297, 126635.

  This article focuses on EPR policies in China, including EPR for automobiles. It discusses the progress made in implementing EPR policies, the challenges faced, and policy implications for the future.

EPR for automobile manufacturers is an important policy approach for promoting sustainable end-of-life management of automobiles and reducing waste in the automobile sector.

8.2.4 Case Studies

- **Examples of green engineering practices in the automobile sector**

Here are some examples of green engineering practices in the automobile sector:

- ➢ Light weighting: The use of lightweight materials in automobile manufacturing, such as aluminum and carbon fiber, can improve fuel efficiency and reduce emissions [36].
- ➢ Hybrid and electric vehicles: The development and production of hybrid and electric vehicles can significantly reduce emissions and improve fuel efficiency [34].
- ➢ Sustainable manufacturing: The implementation of sustainable manufacturing practices, such as lean manufacturing, can reduce waste and improve resource efficiency in automobile manufacturing [35].
- ➢ End-of-life management: Effective end-of-life management practices, such as recycling and repurposing, can reduce waste and increase the recovery of valuable materials from automobiles [32].
- ➢ Eco-driving: The promotion of eco-driving practices, such as reducing speed and avoiding harsh acceleration and braking, can improve fuel efficiency and reduce emissions [33].

These are just a few examples of the many green engineering practices that are being implemented in the automobile sector to improve sustainability and reduce environmental impacts.

- **Success stories of companies implementing green engineering in automobiles**

Here are some success stories of companies implementing green engineering in automobiles:

- ➢ Tesla: Tesla is a company that is known for its commitment to sustainability and green engineering practices. The company's electric vehicles are designed to be energy-efficient and to have minimal environmental impact [40].
- ➢ Toyota: Toyota has been a leader in hybrid vehicle technology, with the Prius being one of the first commercially successful hybrid vehicles. The company has also implemented sustainable manufacturing practices and is committed to reducing the environmental impact of its operations [37].

- BMW: BMW has implemented sustainable manufacturing practices, such as reducing waste and water consumption, in its production facilities. The company has also developed electric and hybrid vehicles and is working to reduce the emissions of its gasoline and diesel vehicles [39].
- Ford: Ford has implemented sustainable manufacturing practices, such as reducing energy and water consumption and increasing the use of renewable energy in its facilities. The company has also developed hybrid and electric vehicles and is committed to reducing the emissions of its gasoline and diesel vehicles [41].
- Volvo: Volvo has made a commitment to electrify its entire vehicle lineup by 2030 and has implemented sustainable manufacturing practices in its production facilities. The company is also exploring the use of recycled materials in its vehicles [38].

These companies are just a few examples of the many automobile manufacturers that are implementing green engineering practices to improve sustainability and reduce environmental impacts.

- **The impact of green engineering on the automobile sector**

  - Green engineering has had a significant impact on the automobile sector, both in terms of environmental sustainability and economic benefits. The implementation of green engineering practices has helped the industry to reduce its environmental footprint and improve resource efficiency. It has also led to the development of new technologies and innovation in the sector, which has led to increased competitiveness and market share. Some studies have shown that the adoption of green engineering practices has led to a reduction in energy consumption and greenhouse gas emissions in the sector.
  - For example, a study by the International Council on Clean Transportation found that the implementation of greenhouse gas emissions standards in the US led to a reduction of 590 million metric tons of carbon dioxide equivalent emissions between 2012 and 2017. Another study by the European Automobile Manufacturers Association found

that the adoption of lightweight materials in car manufacturing could lead to a 6% reduction in fuel consumption and a corresponding reduction in CO_2 emissions [42].

- ➢ Overall, green engineering has become increasingly important in the automobile sector as companies seek to reduce their environmental impact and meet regulatory requirements. It has also become a key driver of innovation and competitiveness in the sector, with companies investing heavily in research and development of new technologies and materials.

8.3 Conclusion

Green engineering has become increasingly important in the automobile sector, as the industry faces pressure to reduce its environmental impact and meet regulatory requirements. The adoption of green engineering practices has helped the industry to reduce its carbon footprint and improve resource efficiency, while also driving innovation and competitiveness in the sector. This has led to the development of new technologies and materials, such as lightweight materials, hybrid and electric powertrains, and advanced driver assistance systems.

While green engineering has already made significant progress in the automobile sector, there is still much work to be done to achieve truly sustainable mobility. This will require continued investment in research and development, as well as collaboration between stakeholders across the value chain, including automobile manufacturers, suppliers, policymakers, and consumers.

The future of the automobile sector lies in the continued adoption of green engineering practices and the development of innovative technologies that can drive sustainability and competitiveness in the industry.

- **Future directions for green engineering in the automobile sector**

As the automobile sector continues to evolve and face new challenges, green engineering will play an increasingly important role in shaping its future. Some potential future directions for green engineering in the automobile sector include:

- ➢ Advancements in electric and autonomous vehicles: Electric vehicles are becoming increasingly popular as a sustainable alternative to traditional internal combustion engine vehicles, and the development of autonomous driving technology could further reduce the environmental impact of automobiles by improving fuel efficiency and reducing emissions.

- Integration of renewable energy sources: The integration of renewable energy sources, such as solar and wind power, into the design and operation of automobiles could help to further reduce their carbon footprint and improve their sustainability.
- Development of circular economy models: The implementation of circular economy models, such as closed-loop recycling and remanufacturing, could help to reduce waste and extend the lifecycle of automobiles, making them more sustainable and resource-efficient.
- Focus on sustainable mobility solutions: The development of sustainable mobility solutions, such as shared mobility and public transportation, could help to reduce the number of vehicles on the road and improve overall resource efficiency in the automobile sector.
- Adoption of green chemistry principles: The adoption of green chemistry principles in the design and manufacturing of automobile materials and components could further improve their sustainability and reduce their environmental impact.

The future of green engineering in the automobile sector will involve continued innovation, collaboration, and a focus on sustainability and resource efficiency.

- **The potential benefits of continued implementation of green engineering in automobiles.**

The continued implementation of green engineering in automobiles has the potential to bring about a wide range of benefits, including:

- Reduced environmental impact: Green engineering practices can help to reduce the environmental impact of automobiles by reducing emissions, improving fuel efficiency, and promoting sustainable manufacturing and end-of-life practices.
- Improved resource efficiency: The adoption of green engineering practices can help to improve resource efficiency

in the automobile sector by promoting the use of recycled and sustainable materials, reducing waste, and extending the lifecycle of vehicles.

- ➢ Cost savings: Green engineering practices can help to reduce costs associated with manufacturing, operations, and end-of-life processes, while also providing opportunities for new revenue streams through the development of sustainable products and services.
- ➢ Increased competitiveness: Companies that adopt green engineering practices can gain a competitive advantage by meeting consumer demand for sustainable products and services, reducing their environmental impact, and improving their brand image.
- ➢ Enhanced innovation: The adoption of green engineering practices can drive innovation in the automobile sector by encouraging the development of new materials, technologies, and business models that promote sustainability and resource efficiency.

The continued implementation of green engineering in automobiles can help to promote a more sustainable and resilient automobile sector, while also creating new opportunities for growth and innovation.

References

1. Anastas, P.T. and Zimmerman, J.B., Design through the 12 Principles of Green Engineering, in: *Environmental Science and Technology*, vol. 37, pp. 94A–101A, 2003.
2. Jaramillo, P., Kahn Ribeiro, S., Newman, P., Dhar, S., Diemuodeke, O.E., Kajino, T., Lee, D.S. *et al.*, Transport (Chapter 10), 1049–1160, 2022.
3. EPA, U.S. Environmental Protection Agency, 2020, Retrieved from https://www.epa.gov/greenengineering.
4. Pacheco-Torgal, F., Labrincha, J.A., Cabeza, L.F., *Handbook of recycled concrete and demolition waste*, Woodhead Publishing, Cambridge, United Kingdom, 2013.
5. National Renewable Energy Laboratory, Electric Vehicle Life-Cycle Emissions and Environmental Footprint, 2020, Retrieved from https://www.nrel.gov/docs/fy20osti/73431.pdf.

6. Sinha, A. and Pal, S.K., Lightweight Materials for Automotive Sector, in: *Lightweight Materials for Sustainable Automotive Engineering*, pp. 3–22, Springer, Cham, Switzerland, 2018.
7. Gawel, A. and Kocot, P., The use of bioplastics in the automotive industry. *J. KONES Powertrain Transport*, 22, 2, 119–125, 2015.
8. Booth, K., Lean manufacturing: the green approach. *Assembly Mag.*, 60, 7, 14–15, 2017.
9. Fehrenbacher, K., *The road to a green auto industry*, GreenBiz, 2019, Retrieved from https://www.greenbiz.com/article/road-green-auto-industry.
10. Ahmed, S., Waqar, M.M., Kim, D.S., Life cycle assessment of electric vehicles: a review. *J. Clean. Prod.*, 242, 118535, 2020.
11. Bocken, N.M., Short, S.W., Rana, P., Evans, S., , and practice review to develop sustainable business model archetypes. *J. Clean. Prod.*, 198, 891–901, 2019.
12. Brown, T. and Cherrett, T., Zero waste manufacturing–a review and challenges. *Int. J. Sustain. Eng.*, 7, 1, 2–10, 2014.
13. Mulej, M., Potocan, V., Sinkovic, T., The adoption of sustainable practices in the automotive industry. *Sustainability*, 11, 4, 1173, 2019.
14. Ferretti, I., Zanoni, S., Pilati, F., Saccani, N., The impact of lean manufacturing on environmental sustainability: a conceptual framework and research agenda. *J. Clean. Prod.*, 226, 959–969, 2019.
15. Ramanathan, R., Lean manufacturing and environmental sustainability: An exploratory analysis. *J. Clean. Prod.*, 172, 4088–4103, 2018.
16. Zhang, X., Wang, S., Ren, Y., Liu, X., Hu, S.J., Design for lean and green manufacturing: A review. *J. Clean. Prod.*, 222, 625–638, 2019.
17. Rashed, M.M., Fathabadi, H., AbouRizk, S., A review of green practices in transportation systems: An operational perspective. *J. Clean. Prod.*, 183, 1196–1212, 2018.
18. Yildiz, A. and Ercan, T., Green driving: Strategies, challenges, and future directions. *Transp. Res. Part D Transport Environ.*, 55, 307–320, 2017.
19. Hoffman, T.A., Wang, Y., Lee, H., Assanis, D.N., Aerodynamics and vehicle design for reduced fuel consumption. *Prog. Aerosp. Sci.*, 106, 1–28, 2019.
20. Li, S., Lv, J., Xie, Y., Li, G., Lightweight materials for automotive applications: A review. *J. Mater. Eng. Perform.*, 28, 11, 6724–6738, 2019.
21. Mamat, R., Abdullah, A.Z., Yusop, A.F., Najafi, G., Yusaf, T., A review on the effects of engine design parameters, fuels and lubricants on piston ring pack friction and oil consumption. *Renew. Sustain. Energy Rev.*, 94, 917–939, 2018.
22. Zhang, S., Xu, X., Li, Y., Sun, X., Review of intelligent transportation systems: towards improving transportation sustainability. *J. Clean. Prod.*, 185, 476–489, 2018.
23. Chauhan, A., Mishra, A.K., Srivastava, S.K., Electric vehicles and renewable energy integration for sustainable development. *Energy Procedia*, 149, 139–146, 2018.

24. Gul, M., Alam, M.S., Rasheed, A.K., The impact of electric vehicles on reducing greenhouse gas emissions and energy conservation. *Renew. Sustain. Energy Rev.*, 119, 109606, 2020.
25. Zahedi, A., Mikulčić, H., Filipović, D., Duić, N., Electric vehicle as a game changer in energy and environmental sustainability. *Renew. Sustain. Energy Rev.*, 135, 110191, 2021.
26. Alvarez, J.C., González-Bravo, R., Ortega, J., Alternative fuels and engine efficiency improvement as strategies to reduce CO2 emissions in the road transport sector. *Energies*, 13, 17, 4264, 2020.
27. Kaiser, M. and Tran, M., Lightweight materials in the automotive industry. *J. Mater. Res. Technol.*, 9, 4, 8838–8851, 2020.
28. Kirchain, R.E., Olivetti, E.A., Hawkins, T.R., Design for end-of-life in the automotive industry. *J. Clean. Prod.*, 196, 698–713, 2018.
29. Li, J., Chen, H., Zuo, T., Huang, Y., End-of-life vehicle management in China: Status quo, challenges and opportunities. *Waste Manage. (Oxford)*, 85, 356–364, 2019.
30. Nabavi, S.R., End-of-life vehicles: Environmental and health impacts and sustainable solutions. *J. Clean. Prod.*, 134, 45–55, 2016.
31. UNEP, Extended producer responsibility: A guidance manual for governments, 2018, Retrieved from https://www.unenvironment.org/resources/report/extended-producer-responsibility-guidance-manual-governments.
32. Sorrell, K., Schäfer, A., Laing, R., The end-of-life management of electric vehicles: A review. *Renew. Sustain. Energy Rev.*, 115, 109370, 2019.
33. Filimonau, V., Gherbin, A., Matute, J., Eco-driving: A review of research on its environmental and economic benefits. *J. Clean. Prod.*, 280, 124247, 2021.
34. Gupta, S., Ahmed, S., Tiwari, G., Environmental sustainability and business prospects of hybrid and electric vehicles. *J. Clean. Prod.*, 289, 125619, 2021.
35. Shahbaz, M., Zafar, S., Zaman, K., Lean manufacturing and sustainability: A review and future research agenda. *J. Clean. Prod.*, 290, 125915, 2021.
36. Yang, X., Jiang, H., Wang, L., A review of lightweight design of automobiles for fuel economy improvement and CO2 reduction. *J. Clean. Prod.*, 308, 127153, 2021.
37. Krause, F., Stock, H., Hohmann, M., Toyota's sustainable business operations: A strategic analysis. *Sustainability*, 13, 4, 1866, 2021.
38. Kuo, T.C., Kuo, H.H., Chiu, A.S.F., A case study of the green supply chain management of Volvo Car Corporation. *J. Clean. Prod.*, 311, 127248, 2021.
39. Lindberg, E., Alm, H., Jönsson, A., Environmental assessment of BMW's electric i3: Life cycle perspective on manufacturing and recycling. *J. Clean. Prod.*, 275, 123155, 2020.
40. Liu, B., Yu, Q., Wang, X., The sustainable development of Tesla Inc.: From perspective of green technology innovation. *J. Clean. Prod.*, 280, 124546, 2021.
41. Shen, Z., Jiang, Z., Zuo, J., Sustainability evaluation of Ford Motor Company's sustainable manufacturing practices. *J. Clean. Prod.*, 279, 123719, 2021.

42. International Council on Clean Transportation, Fuel economy and greenhouse gas emissions standards for passenger cars and light trucks, 2018, Retrieved from https://theicct.org/sites/default/files/publications/US-LD-fuel-economy-GHG-standards_ICCT-White-Paper_24072018.pdf.
43. European Automobile Manufacturers Association, Fuel efficiency: Technology options for the EU passenger car fleet, 2013, https://www.acea.be/uploads/publications/ACEA_Policy_Report_Fuel_Efficiency.pdf.

9

Towards Sustainable Manufacturing: Integrating Digital Technologies on the Green Path

Harjit Singh*, Mukhtiar Singh, Jasvinder Singh and Maninder Singh

School of Mechanical Engineering, Lovely Professional University, Punjab, India

Abstract

This book chapter discusses the integration of digital technologies and green manufacturing practices towards sustainable manufacturing. The chapter provides a definition and importance of sustainable manufacturing, challenges to achieving sustainable manufacturing, and an overview of digital technologies for sustainable manufacturing. Additionally, it discusses green manufacturing practices, such as lean manufacturing, energy efficiency, waste reduction and recycling, and sustainable supply chain management. The chapter also explores the integration of digital technologies and green manufacturing practices and the challenges to achieving this integration. The implications for practice and research and recommendations for future work in sustainable manufacturing are also discussed. The chapter concludes by emphasizing the significance of integrating digital technologies and green manufacturing practices towards sustainable manufacturing for a better future.

Keywords: Lean manufacturing, energy efficiency, waste reduction, sustainable supply chain management, energy consumption

9.1 Introduction

The need for sustainable manufacturing is becoming increasingly important due to the growing concerns about the environmental impact of indus

**Corresponding author*: Harjit.17005@lpu.com.in

Chandan Deep Singh, Talwinder Singh and Davinder Singh (eds.) Digitization and Manufacturing Performance: An Environmental Perspective, (235–264) © 2025 Scrivener Publishing LLC

trial activities. In recent years, there has been a growing body of research highlighting the negative environmental consequences of traditional manufacturing practices, such as increased carbon emissions, energy consumption, and waste generation [3, 4].

Sustainable manufacturing is a holistic approach to manufacturing that seeks to reduce the environmental impact of industrial activities while also improving efficiency and profitability. Sustainable manufacturing involves a range of practices and technologies that aim to reduce the use of resources, minimize waste, and promote a circular economy [1].

The importance of sustainable manufacturing has been recognized by a number of international organizations and initiatives, including the United Nations Sustainable Development Goals (SDGs) and the European Union's Circular Economy Action Plan (CEAP). The SDGs aim to promote sustainable development by addressing a range of global challenges, including climate change, poverty, and inequality. The CEAP is a comprehensive policy framework that aims to promote a more circular and resource-efficient economy in Europe by reducing waste and increasing resource productivity [2].

In addition to its environmental benefits, sustainable manufacturing can also lead to economic and social benefits. For example, sustainable manufacturing practices can reduce operational costs by increasing efficiency and reducing waste [5]. Furthermore, sustainable manufacturing can help companies to meet the growing demand for environmentally-friendly products and services, which can enhance their reputation and market position [6].

Overall, the need for sustainable manufacturing is clear, and it is increasingly being recognized by governments, organizations, and industry stakeholders as an important strategy for addressing global environmental challenges while also promoting economic and social development.

9.1.1 Sustainable Manufacturing

Sustainable manufacturing is an approach to manufacturing that seeks to minimize negative environmental impacts while maintaining economic competitiveness and social well-being. The following are some definitions of sustainable manufacturing from scholarly sources:

"Sustainable manufacturing is the creation of manufactured products that use processes that are non-polluting, conserve energy and natural resources, and are economically sound and safe for employees, communities, and consumers" [8].

"Sustainable manufacturing is the creation of products through economically-sound processes that minimize negative environmental impacts while conserving energy and natural resources, and providing safe and healthy working conditions for employees" [10].

"Sustainable manufacturing is the design, development and implementation of manufacturing systems that are ecologically" [7].

"Sustainable manufacturing can be defined as the creation of products in a way that minimizes negative environmental impacts, conserves energy and natural resources, promotes social responsibility, and is economically feasible" [9].

These definitions highlight the three pillars of sustainability - environmental, economic, and social—and emphasize the importance of considering all three in manufacturing processes. Sustainable manufacturing aims to minimize negative environmental impacts by reducing the use of resources, minimizing waste, and promoting circular economy principles. At the same time, it seeks to maintain economic competitiveness by improving efficiency and reducing costs, and promoting social well-being by ensuring safe and healthy working conditions for employees and contributing to the communities in which it operates.

9.1.2 Importance of Sustainable Manufacturing

Sustainable manufacturing is becoming increasingly important due to a number of environmental, social, and economic challenges facing society. The following are some of the key reasons why sustainable manufacturing is important.

Environmental challenges: Sustainable manufacturing is important in order to minimize negative environmental impacts of manufacturing processes, such as resource depletion, pollution, and greenhouse gas emissions. According to a study by researchers [9], sustainable manufacturing practices can help to reduce environmental impacts by promoting the efficient use of resources, waste reduction, and the adoption of renewable energy sources.

Resource scarcity: With finite resources and increasing demand for products, sustainable manufacturing practices can help to reduce the use of resources and promote circular economy principles. According to scientist ""[7], sustainable manufacturing can promote the efficient use of resources by optimizing processes, reducing waste, and promoting recycling and reuse.

Social responsibility: Sustainable manufacturing is important in order to promote social responsibility and ensure safe and healthy working conditions for employees. According to studies [10] sustainable manufacturing practices can promote social well-being by ensuring fair labor practices, promoting employee health and safety, and contributing to the communities in which companies operate.
Economic competitiveness: Sustainable manufacturing can promote economic competitiveness by reducing costs, improving efficiency, and promoting innovation. Sustainable manufacturing practices can reduce costs by minimizing waste, improving resource efficiency, and reducing environmental risks [8].

Overall, sustainable manufacturing is important in order to minimize negative environmental impacts, promote social well-being, and maintain economic competitiveness in the face of global challenges.

9.1.3 Challenges to Achieving Sustainable Manufacturing

Despite the importance of sustainable manufacturing, there are a number of challenges that must be addressed in order to achieve sustainable production practices. The following are some of the key challenges, supported by citations from scholarly sources:

Complex supply chains: Sustainable manufacturing requires collaboration across the entire supply chain, which can be difficult to achieve due to the complexity of modern supply chains. Sustainable manufacturing requires coordinated action across different organizations, which can be difficult to achieve in practice [12].
Technical barriers: Sustainable manufacturing can require significant investments in new technologies and processes, which can be difficult for some companies to implement. The adoption of sustainable manufacturing practices can be hindered by technical barriers, such as high costs and the need for specialized knowledge [13].
Lack of standards: The lack of widely accepted standards and metrics for sustainable manufacturing can make it difficult to measure and compare sustainability performance across different organizations. The lack of standardization can make it difficult for companies to communicate their sustainability performance to stakeholders [11].
Regulatory frameworks: While regulations can help to promote sustainable manufacturing practices, they can also create barriers to adoption due to compliance costs and other issues. Regulations can create challenges for

sustainable manufacturing by imposing additional costs on companies and limiting their flexibility to adopt new technologies and processes [14].

Overall, achieving sustainable manufacturing requires addressing a range of technical, organizational, and regulatory challenges, and promoting collaboration across the supply chain. Addressing these challenges will require significant effort from companies, policymakers, and other stakeholders.

9.1.4 Digital Technologies for Sustainable Manufacturing

Digital technologies are transforming the manufacturing industry, enabling more efficient and sustainable production processes. The following is an overview of some of the key digital technologies for manufacturing:

Additive Manufacturing: Also known as 3D printing, additive manufacturing involves creating three-dimensional objects by adding layers of material. Additive manufacturing has the potential to reduce material waste, enable more complex geometries, and provide faster prototyping compared to traditional manufacturing methods [15].
Robotics and Automation: Robotics and automation technologies can enable greater precision and efficiency in manufacturing processes. Robotics and automation can also improve worker safety and reduce errors in manufacturing processes [16].
Internet of Things (IoT): IoT technologies enable connected devices to share data and communicate with each other. IoT can enable more efficient supply chain management, real-time monitoring of production processes, and improved quality control [17].
Big Data Analytics: Big data analytics can help manufacturers to identify patterns and insights in large datasets, enabling more informed decision-making. Big data analytics can also help to optimize production processes, reduce waste, and improve product quality [19].
Virtual and Augmented Reality: Virtual and augmented reality technologies can enable manufacturers to simulate production processes and visualize products in 3D. These technologies can improve product design, reduce prototyping costs, and enable remote collaboration [18].

Overall, digital technologies are transforming manufacturing by enabling more efficient, sustainable, and flexible production processes. The adoption of these technologies can help manufacturers to remain competitive and meet the evolving needs of consumers and other stakeholders.

9.1.5 Advantages of Digital Technologies for Sustainable Manufacturing

Digital technologies offer several advantages for sustainable manufacturing, including reduced waste, increased efficiency, and improved product quality. The following are some of the advantages of digital technologies for sustainable manufacturing:

Reduced Waste: Digital technologies, such as additive manufacturing, can significantly reduce waste by enabling the creation of complex geometries and reducing the amount of material needed for production. Additive manufacturing has the potential to reduce material waste by up to 80% compared to traditional manufacturing methods [21].
Increased Efficiency: Digital technologies can improve efficiency in manufacturing processes by enabling real-time monitoring and control of production processes. IoT technologies can enable more efficient energy use and reduce downtime in manufacturing processes, resulting in significant cost savings [20].
Improved Product Quality: Digital technologies, such as big data analytics, can help to identify quality issues early in the production process and enable more informed decision-making. Big data analytics can improve product quality by identifying patterns and insights in production data [19].
Greater Flexibility: Digital technologies can enable greater flexibility in manufacturing processes, allowing for more customized and adaptable production processes. Additive manufacturing can enable the creation of customized parts without the need for specialized tooling or equipment [15].
Enhanced Sustainability: Digital technologies can help to reduce the environmental impact of manufacturing processes by enabling more sustainable practices. Robotics and automation can reduce energy use and waste in manufacturing processes, while IoT technologies can enable more sustainable supply chain management [16].

Overall, digital technologies offer several advantages for sustainable manufacturing, enabling more efficient, flexible, and sustainable production processes.

9.2 Digital Technologies for Sustainable Manufacturing with Internet of Things (IoT)

The Internet of Things (IoT) is a digital technology that connects physical devices to the internet, enabling real-time monitoring and control of manufacturing processes. IoT can have several advantages for sustainable manufacturing, including improved energy efficiency, reduced waste, and enhanced product quality. The following are some examples of how IoT can support sustainable manufacturing:

Energy Efficiency: IoT can enable more efficient energy use in manufacturing processes by enabling real-time monitoring and control of energy consumption. IoT can enable the identification of energy inefficiencies in manufacturing processes and enable more targeted energy management strategies, resulting in significant energy savings [20].
Waste Reduction: IoT can enable the reduction of waste in manufacturing processes by enabling real-time monitoring and control of production processes. IoT can enable the identification of waste in production processes and enable more efficient use of materials, resulting in reduced waste [21].
Product Quality: IoT can improve product quality by enabling real-time monitoring and control of production processes. IoT can enable the identification of quality issues early in the production process and enable more informed decision-making, resulting in improved product quality [19].
Supply Chain Management: IoT can enable more sustainable supply chain management by enabling real-time tracking of products and materials. IoT can enable more efficient and sustainable supply chain management by enabling real-time tracking of products and materials, resulting in reduced waste and improved efficiency [16].

Overall, IoT can support sustainable manufacturing by enabling more efficient, flexible, and sustainable production processes.

9.3 Digital Technologies for Sustainable Manufacturing with Artificial Intelligence

Artificial intelligence (AI) is a digital technology that can have several advantages for sustainable manufacturing. AI can enable more efficient and sustainable production processes, reduce waste, and improve product quality. The following are some examples of how AI can support sustainable manufacturing, supported by citations from scholarly sources:

Energy efficiency: AI can enable more efficient energy use in manufacturing processes by enabling real-time optimization of energy consumption. AI can enable more efficient energy use by optimizing production processes and reducing energy waste, resulting in significant energy savings [24].
Waste reduction: AI can enable the reduction of waste in manufacturing processes by enabling real-time monitoring and control of production processes. AI can enable the identification of waste in production processes and enable more efficient use of materials, resulting in reduced waste [25].
Product quality: AI can improve product quality by enabling real-time monitoring and control of production processes. AI can enable the identification of quality issues early in the production process and enable more informed decision-making, resulting in improved product quality [22].
Supply chain management: AI can enable more sustainable supply chain management by enabling real-time tracking of products and materials. AI can enable more efficient and sustainable supply chain management by enabling real-time tracking of products and materials, resulting in reduced waste and improved efficiency [23].

Overall, AI can support sustainable manufacturing by enabling more efficient, flexible, and sustainable production processes.

9.4 Digital Technologies for Sustainable Manufacturing with Digital Twins

Digital twins are a digital technology that can have several advantages for sustainable manufacturing. Digital twins are virtual models of physical assets, processes, and systems that can be used for simulation, analysis, and optimization. The following are some examples of how digital twins can support sustainable manufacturing:

Energy efficiency: Digital twins can enable more efficient energy use in manufacturing processes by enabling simulation and optimization of production processes. Digital twins can enable more efficient energy use by optimizing production processes and reducing energy waste, resulting in significant energy savings [26].

Waste reduction: Digital twins can enable the reduction of waste in manufacturing processes by enabling simulation and optimization of production processes. Digital twins can enable the identification of waste in production processes and enable more efficient use of materials, resulting in reduced waste [29].
Product quality: Digital twins can improve product quality by enabling simulation and optimization of production processes. Digital twins can enable the identification of quality issues early in the production process and enable more informed decision-making, resulting in improved product quality [27].
Maintenance and repair: Digital twins can enable more sustainable maintenance and repair of manufacturing assets by enabling real-time monitoring and analysis of asset performance. Digital twins can enable more efficient and sustainable maintenance and repair by enabling real-time monitoring and analysis of asset performance, resulting in reduced downtime and maintenance costs [28].

Digital twins can support sustainable manufacturing by enabling more efficient, flexible, and sustainable production processes.

9.5 Digital Technologies for Sustainable Manufacturing with Additive Manufacturing (3D Printing)

Additive manufacturing, also known as 3D printing, is a digital technology that can support sustainable manufacturing in several ways. The following are some examples of how additive manufacturing can contribute to sustainable manufacturing:

Waste reduction: Additive manufacturing can reduce waste in manufacturing processes by enabling more efficient use of materials. Additive manufacturing can enable more efficient use of materials by enabling the production of complex geometries with minimal waste [28].
Resource efficiency: Additive manufacturing can increase resource efficiency by enabling the production of parts with less material and energy consumption. Additive manufacturing can enable the production of lightweight parts that require less energy and materials, resulting in reduced environmental impact [31].
Design optimization: Additive manufacturing can enable the optimization of product design for sustainability by enabling the production of complex

geometries and lightweight structures. Additive manufacturing can enable the optimization of product design for sustainability by enabling the production of complex geometries and lightweight structures, resulting in improved energy and material efficiency [30].
Localized production: Additive manufacturing can enable localized production, reducing the need for transportation and associated carbon emissions. Additive manufacturing can enable the production of customized products at the point of use, reducing the need for transportation and associated carbon emissions [32, 33].

Additive manufacturing can contribute to sustainable manufacturing by enabling more efficient use of materials and energy, optimized product design, and localized production.

9.6 Digital Technologies for Sustainable Manufacturing with Augmented Reality (AR)

Augmented reality (AR) is a digital technology that allows the enhancement of a real-world environment with virtual objects or information. AR can support sustainable manufacturing in various ways:

Enhancing worker efficiency: AR can enhance worker efficiency by providing real-time information and instructions. AR-based work instructions can reduce assembly time and error rates, improving worker productivity and reducing waste [35].
Reducing energy consumption: AR can reduce energy consumption by enabling the visualization of energy usage and the optimization of energy consumption. AR can enable the visualization of energy usage and provide real-time feedback to reduce energy consumption and improve energy efficiency [36].
Improving maintenance: AR can improve maintenance by enabling remote monitoring and visualization of machinery and equipment. For example, AR-based remote maintenance can improve the efficiency of maintenance tasks and reduce the need for onsite maintenance visits, resulting in reduced transportation-related carbon emissions [37].
Supporting sustainable design: AR can support sustainable design by enabling the visualization of products and their environmental impact. AR can enable the visualization of products in different environments and

provide environmental impact information, enabling the identification of design improvements to reduce environmental impact [34].

AR can support sustainable manufacturing by enhancing worker efficiency, reducing energy consumption, improving maintenance, and supporting sustainable design.

9.7 Green Path for Sustainable Manufacturing

The green path for sustainable manufacturing involves the integration of various technologies and strategies to minimize the environmental impact of manufacturing processes and products. Here are some examples of green paths for sustainable manufacturing:

Renewable energy: The use of renewable energy sources such as solar, wind, and hydropower can reduce greenhouse gas emissions and minimize the environmental impact of manufacturing. The adoption of renewable energy in manufacturing can also reduce the dependence on fossil fuels and enhance energy security [39].
Closed-loop systems: Closed-loop systems involve the recycling and reuse of materials and waste streams within the manufacturing process. This can reduce resource consumption and waste generation, as well as minimize the environmental impact of the manufacturing process. Closed-loop systems can also reduce the cost of raw materials and waste disposal [40].
Life cycle assessment: Life cycle assessment (LCA) is a tool used to assess the environmental impact of a product or process throughout its entire life cycle. The use of LCA can support sustainable manufacturing by identifying areas for improvement in terms of environmental impact. LCA can also support decision-making for sustainable manufacturing by comparing the environmental impact of different manufacturing processes and product designs [41].
Circular economy: The circular economy is a model that emphasizes the reuse and regeneration of materials and products, rather than their disposal. The circular economy can support sustainable manufacturing by minimizing waste and reducing the consumption of virgin materials. The circular economy can also provide economic benefits by reducing the cost of raw materials and waste disposal [38].

The green path for sustainable manufacturing involves the adoption of renewable energy, closed-loop systems, life cycle assessment, and circular economy models to minimize the environmental impact of manufacturing processes and products.

9.8 Introduction to Green Manufacturing

Green manufacturing refers to the use of environmentally friendly manufacturing processes and practices that minimize waste, reduce pollution, and conserve resources. It involves integrating sustainable strategies into all aspects of manufacturing, from product design to disposal. Here are some definitions of green manufacturing:

Green manufacturing is "a process of integrating environmental considerations into the design and development of products and processes, with the aim of minimizing environmental impact and optimizing resource efficiency throughout the product life cycle" [44].

The United Nations Industrial Development Organization (UNIDO) defines green manufacturing as "a manufacturing model that minimizes the use of natural resources and the generation of waste, while maximizing economic efficiency and minimizing environmental impact" [43]. The European Commission (2014) defines green manufacturing as "a production process that is environmentally sound throughout its entire life cycle, from the extraction of raw materials to the final disposal of the product." Green manufacturing involves the adoption of sustainable practices and technologies to reduce the environmental impact of manufacturing processes and products, while also improving economic efficiency [42].

9.8.1 Benefits of Green Manufacturing

Green manufacturing offers a range of benefits to companies, the environment, and society as a whole. Here are some of the benefits of green manufacturing:

Cost savings: Green manufacturing can help companies reduce their operating costs by using fewer resources, generating less waste, and improving energy efficiency. Studies have shown that companies that adopt sustainable manufacturing practices can achieve significant cost savings [45, 46].
Improved environmental performance: Green manufacturing can help reduce the environmental impact of manufacturing processes and products. By using sustainable practices and technologies, companies can

minimize their carbon footprint, reduce water and energy consumption, and decrease waste generation [48, 49].

Enhanced reputation and customer loyalty: Companies that adopt green manufacturing practices are often seen as socially responsible and environmentally conscious. This can enhance their reputation and attract environmentally conscious customers, which can lead to increased sales and customer loyalty [47].

Regulatory compliance: Green manufacturing can help companies comply with environmental regulations and avoid fines and penalties for non-compliance [50].

Innovation and competitiveness: Green manufacturing can spur innovation and competitiveness by encouraging companies to develop new and sustainable products, processes, and technologies. This can create new market opportunities and increase the company's competitiveness [45, 48].

Green manufacturing offers a range of benefits to companies, society, and the environment, making it a compelling strategy for sustainable development.

9.8.2 Green Manufacturing Practices with Lean Manufacturing

Lean manufacturing is a methodology aimed at eliminating waste and improving efficiency in manufacturing processes. It involves continuous improvement of processes and the elimination of non-value-adding activities. Lean manufacturing is compatible with green manufacturing practices as both aims to reduce waste and improve efficiency.

Studies have shown that the implementation of lean manufacturing practices can lead to significant reductions in environmental impact. For example, the implementation of lean manufacturing practices resulted in a reduction in waste generation and energy consumption in a manufacturing plant. Similarly, the implementation of lean manufacturing practices in a furniture manufacturing company resulted in a reduction in material waste, energy consumption, and greenhouse gas emissions. Furthermore, lean manufacturing practices can also lead to cost savings for companies. For example, the implementation of lean manufacturing practices in a food packaging company resulted in a reduction in production costs and an improvement in product quality. The integration of lean manufacturing practices with green manufacturing practices can result in a more sustainable and efficient manufacturing process.

9.8.3 Green Manufacturing Practices with Energy Efficiency

Energy efficiency is an important aspect of green manufacturing practices. The following are some examples of green manufacturing practices related to energy efficiency:

Energy auditing: Conducting energy audits is an important green manufacturing practice that helps to identify areas where energy is being wasted and where efficiency can be improved. Energy audits can help manufacturers to identify opportunities to reduce energy consumption and save money. Energy audits can result in energy savings of up to 30%.
Energy-efficient lighting: Switching to energy-efficient lighting can result in significant energy savings. LED lighting, for example, can use up to 80% less energy than traditional incandescent lighting. The widespread adoption of LED lighting could result in energy savings of up to 300 terawatt-hours per year by 2030.
Energy-efficient equipment: Using energy-efficient equipment, such as motors and compressors, can also result in significant energy savings. A study by the American Council for an Energy-Efficient Economy (ACEEE) (2019) found that the use of energy-efficient equipment could result in energy savings of up to 20%.
Renewable energy: Using renewable energy sources, such as solar or wind power, is another important green manufacturing practice. A study by the National Renewable Energy Laboratory (NREL) (2012) found that the use of renewable energy sources in manufacturing could result in significant energy savings and cost reductions. Incorporating energy-efficient practices into manufacturing processes can result in significant energy savings, cost savings, and environmental benefits.

9.8.4 Green Manufacturing Practices with Waste Reduction and Recycling

Waste reduction and recycling are important aspects of green manufacturing practices. The following are some examples of green manufacturing practices related to waste reduction and recycling:

Material reuse: Reusing materials in manufacturing processes can help to reduce waste and save resources. For example, scrap metal can be melted down and reused in new products. A study by the Environmental Protection Agency (EPA) (2016) found that material reuse can result in significant environmental benefits, including reduced greenhouse gas emissions and resource conservation.

Recycling: Recycling materials, such as paper, plastic, and metal, is another important green manufacturing practice. A study by the EPA (2018) found that recycling can result in significant energy savings and greenhouse gas emissions reductions.

Composting: Composting organic waste, such as food and yard waste can help to reduce waste and create a valuable soil amendment. A study by the EPA (2016) found that composting can result in significant environmental benefits, including reduced greenhouse gas emissions and improved soil health.

Waste reduction: Implementing waste reduction strategies, such as source reduction and process improvements, can help to minimize waste in manufacturing processes. A study by the EPA (2016) found that waste reduction strategies can result in significant environmental benefits, including reduced greenhouse gas emissions and resource conservation. Incorporating waste reduction and recycling practices into manufacturing processes can result in significant environmental benefits, including reduced greenhouse gas emissions, resource conservation, and improved soil health.

9.8.5 Green Manufacturing Practices with Sustainable Supply Chain Management

Sustainable supply chain management is another important aspect of green manufacturing practices. The following are some examples of sustainable supply chain management practices:

Supplier selection: Selecting suppliers who share a commitment to sustainability can help to promote sustainable practices throughout the supply chain. Supplier selection criteria based on environmental and social performance can improve sustainability outcomes in the supply chain.

Green transportation: Choosing transportation methods that minimize environmental impact, such as low-emission vehicles and alternative fuels, is another important aspect of sustainable supply chain management. Green transportation initiatives can result in significant environmental benefits, including reduced greenhouse gas emissions and improved air quality.

Sustainable packaging: Using sustainable packaging materials and designs can help to reduce waste and promote environmental sustainability. A study by the Sustainable Packaging Coalition (2018) found that sustainable

packaging practices can result in significant environmental benefits, including reduced greenhouse gas emissions and resource conservation.
Traceability and transparency: Implementing traceability and transparency measures in the supply chain can help to ensure that sustainable practices are being followed and can improve accountability. Traceability measures can help to promote sustainable practices in the supply chain and improve social and environmental performance [47]. Implementing sustainable supply chain management practices can help to promote environmental sustainability throughout the supply chain and improve social and environmental performance.

9.8.6 Integration of Digital Technologies on the Green Path

The integration of digital technologies in green manufacturing can lead to significant improvements in environmental sustainability and efficiency. The following are some examples of how digital technologies can be integrated into green manufacturing practices:

Internet of Things (IoT): The IoT can be used in green manufacturing to monitor and optimize energy consumption, reduce waste, and improve resource efficiency. IoT can help to reduce energy consumption and greenhouse gas emissions in manufacturing [51].
Artificial Intelligence (AI): AI can be used in green manufacturing to optimize production processes, improve energy efficiency, and reduce waste. AI can help to improve energy efficiency and reduce emissions in manufacturing [52].
Digital Twins: Digital twins can be used in green manufacturing to simulate and optimize production processes, reduce waste, and improve resource efficiency. Digital twins can help to improve resource efficiency and reduce emissions in manufacturing [53].
Additive Manufacturing (3D Printing): Additive manufacturing can be used in green manufacturing to reduce waste and improve resource efficiency by enabling the production of complex parts with minimal material waste. Additive manufacturing can help to reduce waste and improve resource efficiency in manufacturing [54].
Augmented Reality (AR): AR can be used in green manufacturing to improve training and reduce errors, leading to improved efficiency and reduced waste. AR can help to improve training and reduce errors in manufacturing. The integration of digital technologies in green manufacturing can help to reduce waste, improve efficiency, and promote environmental sustainability [55].

9.8.7 Importance of Integrating Digital Technologies and Green Manufacturing Practices

Integrating digital technologies with green manufacturing practices can result in a more sustainable and efficient manufacturing process. Integrating digital technologies such as AI, machine learning, and IoT with sustainable manufacturing practices can lead to a significant reduction in energy consumption and waste generation [58]. Incorporating green manufacturing practices such as waste reduction and recycling, and sustainable supply chain management can improve overall sustainability performance and reduce the environmental impact of manufacturing [57].

Additionally, integrating digital technologies with green manufacturing practices can also lead to cost savings and increased competitiveness. According to a report by the World Economic Forum (2018), companies that adopt sustainable practices and digital technologies can achieve higher levels of efficiency, productivity, and profitability [59]. Integrating digital technologies with lean manufacturing practices can lead to significant improvements in production efficiency and cost reduction [56]. Therefore, it is important to integrate digital technologies and green manufacturing practices to achieve sustainable and efficient manufacturing processes.

9.8.8 Challenges to Integrating Digital Technologies and Green Manufacturing Practices

The integration of digital technologies and green manufacturing practices also comes with some challenges that need to be addressed for successful implementation. These challenges include:

- ➢ Cost: The cost of implementing digital technologies and green manufacturing practices can be high, and some companies may not have the resources to invest in these technologies and practices [63].
- ➢ Skill gap: The implementation of digital technologies and green manufacturing practices requires skilled personnel who can manage the technology and processes effectively. However, there may be a shortage of skilled workers, and training existing employees can be time-consuming and expensive [64].
- ➢ Data privacy and security: The use of digital technologies involves the collection, storage, and analysis of data. This can

pose a risk to data privacy and security if not managed effectively [62].

- Regulatory compliance: The integration of digital technologies and green manufacturing practices may require compliance with new regulations and standards, which can be complex and time-consuming [60].
- Resistance to change: Some employees and stakeholders may resist the implementation of new technologies and practices, which can impede progress and innovation [61].

9.8.9 Case Studies of Successful Integration of Digital Technologies and Green Manufacturing Practices

BMW Group—The BMW Group has implemented a smart and sustainable production approach using digital technologies such as AI, robotics, and IoT, along with green manufacturing practices such as energy-efficient lighting, waste reduction, and renewable energy sources. This has resulted in a 50% reduction in CO_2 emissions per vehicle produced and a 45% reduction in water consumption per vehicle produced since 2006.

Schneider Electric—Schneider Electric has implemented a sustainable manufacturing program that integrates digital technologies such as IoT and AI with green manufacturing practices such as energy efficiency and waste reduction. This has resulted in a 10% reduction in energy consumption, a 12% reduction in water consumption, and a 25% reduction in waste generation per unit of production since 2015.

Siemens—Siemens has implemented a digitalization strategy that integrates digital technologies such as IoT, AI, and digital twins with green manufacturing practices such as energy efficiency and waste reduction. This has resulted in a 20% reduction in CO_2 emissions and a 25% reduction in waste generation since 2014.

Toyota—Toyota has implemented a sustainable manufacturing approach that integrates digital technologies such as IoT and AI with green manufacturing practices such as energy efficiency and waste reduction. This has resulted in a 50% reduction in CO_2 emissions per vehicle produced and a 95% reduction in water consumption per vehicle produced since 2001.

Procter & Gamble—Procter & Gamble has implemented a sustainable manufacturing program that integrates digital technologies such as IoT and AI with green manufacturing practices such as energy efficiency and waste reduction. This has resulted in a 50% reduction in greenhouse gas emissions, a 35% reduction in water consumption, and a 50% reduction in waste generation per unit of production since 2010.

9.9 Future Trends in Sustainable Manufacturing

The field of sustainable manufacturing is constantly evolving, and there are several emerging trends that are expected to shape the future of the industry. Here are some future trends in sustainable manufacturing:

- Circular Economy: The concept of the circular economy is gaining popularity as a sustainable approach to manufacturing. This model focuses on reducing waste and maximizing the use of resources through closed-loop systems, where waste is minimized and materials are reused, recycled or repurposed [65].
- Digital Twins: Digital twins are expected to play a significant role in sustainable manufacturing in the future. By creating a virtual replica of a physical system or product, manufacturers can identify opportunities for optimization, predict performance and improve sustainability throughout the lifecycle of the product [66].
- 5G Connectivity: The next generation of wireless connectivity, 5G, is expected to enable the development of more efficient and sustainable manufacturing systems. The high-speed and low-latency connectivity will allow for real-time monitoring and control of manufacturing processes, improving efficiency and reducing waste [67].
- Additive Manufacturing: Additive manufacturing, also known as 3D printing, has the potential to revolutionize sustainable manufacturing by reducing waste, energy consumption and emissions. The technology enables manufacturers to create products with complex geometries, reducing the need for assembly and minimizing waste material [68].
- Artificial Intelligence: Artificial intelligence (AI) is expected to play a significant role in sustainable manufacturing in the future. By using data and machine learning algorithms, manufacturers can optimize processes, reduce energy consumption and improve sustainability throughout the product lifecycle [69].

9.10 Emerging Digital Technologies for Sustainable Manufacturing

There are several emerging digital technologies that have the potential to enable sustainable manufacturing. Here are some of them:

- Block chain technology: Block chain can be used for supply chain transparency and traceability, allowing companies to track the sustainability of their products and materials [72].
- Augmented reality (AR): AR can be used for training and simulation, reducing the need for physical prototypes and travel, leading to more sustainable manufacturing practices [70].
- Digital twins: Digital twins can enable the optimization of manufacturing processes and product design, resulting in reduced waste and energy consumption [71].
- Robotics and automation: Robotics and automation can improve energy efficiency and reduce waste by enabling more precise and consistent manufacturing processes [83].
- Internet of Things (IoT): IoT can be used for real-time monitoring of manufacturing processes, leading to improved efficiency and reduced waste [73].

9.11 New Trends in Green Manufacturing Practices

Some new trends in green manufacturing practices are:

- Circular economy: A circular economy focuses on reducing waste and promoting resource efficiency by reusing and recycling materials, reducing resource consumption, and extending the life of products. This approach can help to reduce the environmental impact of manufacturing processes while creating new business opportunities [74].
- Sustainable materials: The use of sustainable materials, such as bio-based materials and recycled materials, can help to reduce the environmental impact of manufacturing processes. This trend is gaining popularity as consumers become more concerned about the sustainability of products [78].

- Product life extension: The concept of product life extension involves designing products that can be easily repaired, upgraded, and reused. This approach can help to reduce waste and promote resource efficiency [75].
- Green logistics: Green logistics involves optimizing transportation networks, reducing emissions, and minimizing waste in the supply chain. This approach can help to reduce the environmental impact of manufacturing processes [54, 77].
- Digitalization of green manufacturing practices: The integration of digital technologies, such as the Internet of Things (IoT) and artificial intelligence (AI), with green manufacturing practices can help to optimize processes, reduce waste, and improve energy efficiency [76].

9.12 Future Directions for Sustainable Manufacturing

There are several future directions for sustainable manufacturing, some of which are discussed below:

- Circular economy: The circular economy is an economic system that aims to minimize waste and maximize the use of resources. It is a closed-loop system in which materials are reused, recycled, or regenerated at the end of their useful life. This approach is gaining popularity in sustainable manufacturing as it can reduce waste, energy consumption, and greenhouse gas emissions [81].
- Digitalization of the supply chain: The digitalization of the supply chain can improve transparency and traceability, reduce waste, and optimize the use of resources. Technologies such as block chain, RFID, and IoT can enable better tracking of materials and products throughout the supply chain. This can help identify inefficiencies and reduce waste.
- Biomanufacturing: Biomanufacturing involves using living organisms, such as bacteria and yeast, to produce materials and products. This approach has the potential to reduce the use of non-renewable resources and decrease waste. Biomanufacturing can also produce products that are more environmentally friendly and sustainable [79].

- 3D printing: 3D printing, also known as additive manufacturing, is a technology that can produce complex shapes and designs using less material than traditional manufacturing methods. This can reduce waste and energy consumption. 3D printing can also enable local production, which can reduce the need for transportation and logistics [82].
- Renewable energy: Renewable energy sources, such as solar and wind power, can reduce the carbon footprint of manufacturing operations. These sources can be used to power factories, warehouses, and other facilities, as well as to charge electric vehicles used in the supply chain [80].

9.13 Conclusion

The integration of digital technologies with green manufacturing practices is critical in achieving sustainable manufacturing. Digital technologies such as IoT, AI, digital twins, 3D printing, and AR are increasingly being adopted in manufacturing, leading to improved efficiency, reduced waste, and lower energy consumption. Similarly, green manufacturing practices such as lean manufacturing, energy efficiency, waste reduction, and sustainable supply chain management are being widely embraced in manufacturing, resulting in reduced environmental impact and improved economic performance.

However, the integration of digital technologies with green manufacturing practices faces challenges such as high costs, lack of skilled labor, and resistance to change. Nevertheless, successful case studies demonstrate that the integration of digital technologies with green manufacturing practices can lead to significant benefits, including improved productivity, reduced costs, and enhanced sustainability.

Future trends in sustainable manufacturing are expected to include the adoption of emerging digital technologies such as blockchain, machine learning, and the use of renewable energy sources. Furthermore, new trends in green manufacturing practices may include the circular economy and the use of biomimicry.

The use of digital technologies such as IoT, AI, and digital twins can lead to more efficient and sustainable manufacturing processes, while green manufacturing practices such as waste reduction and sustainable supply chain management can reduce the environmental impact of manufacturing.

However, there are also challenges to the integration of these practices, including issues with data security and privacy, as well as the need for increased collaboration and cooperation between different stakeholders in the manufacturing industry.

Further research is needed to explore the full potential of digital technologies and green manufacturing practices in sustainable manufacturing, including the development of new and innovative solutions that can address these challenges. Additionally, research is needed to explore the social and economic impacts of sustainable manufacturing practices, including the potential for job creation and economic growth.

In summary, the integration of digital technologies with green manufacturing practices provides a path towards sustainable manufacturing, which is critical in addressing the environmental, social, and economic challenges facing the manufacturing sector.

9.14 Future Scope

The following recommendations can be made for future work in sustainable manufacturing:

- Continued exploration of digital technologies: There is a need for further exploration and development of digital technologies to enhance their application in sustainable manufacturing.
- Increased focus on green supply chain management: There is a need for more research and practical solutions that integrate green manufacturing practices into supply chain management.
- Collaboration among stakeholders: Collaboration among stakeholders, including government agencies, industry players, and research institutions, is essential to drive the adoption of sustainable manufacturing practices.
- Development of metrics for measuring sustainability: There is a need to develop metrics that can accurately measure the sustainability performance of manufacturing systems.
- Adoption of circular economy principles: The adoption of circular economy principles in manufacturing can contribute to waste reduction, material recovery, and energy efficiency.

- Training and education: Training and education programs that focus on sustainable manufacturing practices should be developed to enhance the knowledge and skills of manufacturing professionals.
- Policy support: Governments should provide policy support to encourage the adoption of sustainable manufacturing practices, including the development of incentives and regulations.

Overall, there is a need for a holistic approach to sustainable manufacturing that integrates digital technologies and green manufacturing practices. The implementation of these recommendations can contribute to the development of a more sustainable and efficient manufacturing industry.

References

1. Azab, M., Baroutaji, A., El Mhamedi, A., Sustainable Manufacturing: A Literature Review. *J. Manuf. Mater. Process.*, 3, 1, 18, 2019, doi: 10.3390/jmmp3010018.
2. European Commission, Circular Economy Action Plan, 2020, Retrieved from https://ec.europa.eu/environment/circular-economy/index_en.htm.
3. Gbededo, M. A., Liyanage, K., Garza-Reyes, J. A., Towards a Life Cycle Sustainability Analysis: A systematic review of approaches to sustainable manufacturing. *J. Clean. Prod.*, 184, 1002–1015, 2018.
4. Mulligan, C., Bhamra, T., Lockett, A., The Role of Sustainable Manufacturing in Circular Economy. *J. Manuf. Technol. Manage.*, 29, 3, 414–436, 2018, doi: 10.1108/JMTM-08-2017-0121.
5. Sundin, E. and Lindahl, M., Sustainability in Manufacturing: A Review of Recent Developments. *J. Clean. Prod.*, 102, 1–17, 2015, doi: 10.1016/j.jclepro.2014.07.078.
6. Zsidisin, G.A. and Wagner, S.M., Do perceptions become reality? The moderating role of supply chain partners. *J. Bus. Logist.*, 31, 1, 1–25, 2010, doi: 10.1002/j.2158-1592.2010.tb00122.x.
7. Makkar, M. and Singh, S., Sustainable manufacturing: A review. *Int. J. Sustain. Manuf.*, 3, 1, 91–107, 2016, doi: 10.1504/IJSM.2016.077682.
8. Mont, O., Clarifying the concept of product-service system. *J. Clean. Prod.*, 10, 3, 237–245, 2002, doi: 10.1016/S0959-6526(01)00039-0.
9. Silaen, P., Lu, W.F., Shu, C., Wang, Z., Sustainable manufacturing: A review and research agenda. *J. Clean. Prod.*, 220, 133–146, 2019, doi: 10.1016/j.jclepro.2019.02.163.

10. Sundin, E. and Bras, B., Approaches to sustainable manufacturing, in: *Sustainable manufacturing*, S.T. Kumara, J.K.H. Tan, C.M.L. Wu (Eds.), pp. 429–451, Springer, Boston, MA, USA, 2005, doi: 10.1007/0-387-23183-7_18.
11. Azapagic, A., Perdan, S., Clift, R., Sustainable supply chain management in the UK food industry. *J. Clean. Prod.*, 18, 10-11, 1180–1192, 2010, doi: 10.1016/j.jclepro.2010.02.010.
12. Heikkilä, J., Niskanen, E., Ståhle, M., Towards sustainability in manufacturing networks: A systematic literature review. *J. Clean. Prod.*, 270, 122386, 2020, doi: 10.1016/j.jclepro.2020.122386.
13. Kusi-Sarpong, S. and Sarkis, J., A supply chain environmental model for evaluating environmental performance. *Int. J. Prod. Econ.*, 128, 1, 69–76, 2010, doi: 10.1016/j.ijpe.2009.08.033.
14. Zutshi, A., Sohal, A.S., Cavaye, A.L., Sustainable manufacturing: Key research issues, challenges and directions. *Int. J. Prod. Res.*, 51, 23-24, 6580–6598, 2013, doi: 10.1080/00207543.2013.803070.
15. Fathi, M., Dastjerdi, A.V., Vahdatikhaki, F., The role of additive manufacturing in sustainable manufacturing: A review of the literature. *J. Clean. Prod.*, 269, 122381, 2020, doi: 10.1016/j.jclepro.2020.122381.
16. Jovane, F., Westkämper, E., Williams, D., Yan, Y., The future of manufacturing: A new era of opportunity and challenge for the UK. *Int. J. Comput. Integr. Manuf.*, 23, 7, 579–593, 2010, doi: 10.1080/0951192X.2010.498904.
17. Lee, I. and Lee, K., The Internet of Things (IoT): Applications, investments, and challenges for enterprises. *Bus. Horiz.*, 58, 4, 431–440, 2015, doi: 10.1016/j.bushor.2015.03.008.
18. Piyabongkarn, D., Karmacharya, A., Karmacharya, S., Augmented reality and virtual reality in manufacturing: A survey. *J. Ind. Integr. Manage.*, 4, 3, 2050034, 2019, doi: 10.1142/S2424787X20500342.
19. Soltani, R., Azizi, R., Tavakkoli-Moghaddam, R., Big data analytics in manufacturing: A review. *J. Ind. Integr. Manage.*, 5, 1, 2050002, 2020, doi: 10.1142/S2424787X20500028.
20. Wu, Z., Xu, X., Wang, L., A review of the applications of the Internet of Things in logistics and supply chain management. *Int. J. Prod. Res.*, 56, 1-2, 1018–1037, 2018, doi: 10.1080/00207543.2017.1346186.
21. Zhang, Y., Yu, W., Yang, Y., Research progress in reducing waste in additive manufacturing. *Mater. Des.*, 184, 108207, 2019, doi: 10.1016/j.matdes.2019.108207.
22. Chen, X., Wu, M., Jiang, J., Jiang, L., An intelligent quality control system based on machine learning and big data analytics in additive manufacturing. *J. Manuf. Syst.*, 55, 38–47, 2020, doi: 10.1016/j.jmsy.2020.01.002.
23. Gligor, V., Gligor, N., Lascu, D.N., The potential of artificial intelligence for green logistics and supply chain management. *J. Clean. Prod.*, 208, 1040–1051, 2019, doi: 10.1016/j.jclepro.2018.10.272.

24. Tavakoli, S., Azadegan-Mehr, M., Mohammadi-Ivatloo, B., Aslani, A., An AI-based optimization algorithm for energy management in microgrid systems. *J. Energy Storage*, 29, 101304, 2020, doi: 10.1016/j.est.2020.101304.
25. Yang, J., Liu, Y., Qian, Y., Chen, Q., Wang, D., A smart waste reduction approach for additive manufacturing with big data analytics and artificial intelligence. *J. Clean. Prod.*, 221, 102–113, 2019, doi: 10.1016/j.jclepro.2019.02.139.
26. Liu, Y., Zhang, X., Yu, Z., Ma, Y., Wu, J., Energy-efficient production process optimization of forging equipment based on digital twin. *J. Clean. Prod.*, 253, 119908, 2020, doi: 10.1016/j.jclepro.2019.119908.
27. Tao, F., Zuo, Y., Da Xu, L., Zhang, M., Li, Y., Digital twin-driven product design, manufacturing and service with big data. *Int. J. Adv. Manuf. Technol.*, 100, 5-8, 1305–1316, 2019, doi: 10.1007/s00170-018-3150-9.
28. Wang, H., Li, X., Guo, L., Li, Y., Li, J., Predictive maintenance for rotating machinery based on digital twin and deep learning. *J. Intell. Manuf.*, 32, 4, 941–954, 2021, doi: 10.1007/s10845-020-01636-7.
29. Yu, H., Li, H., Tan, R., Wang, J., Huang, G.Q., Multi-objective optimization for reducing waste in process planning with a digital twin. *J. Clean. Prod.*, 254, 120107, 2020, doi: 10.1016/j.jclepro.2019.120107.
30. Chen, H., Zhang, Y., Bai, S., Zhang, L., Li, B., Design for additive manufacturing of sustainable products: A review. *J. Clean. Prod.*, 294, 126020, 2021, doi: 10.1016/j.jclepro.2021.126020.
31. Raja, M.R., Ismail, R., Yusoff, W.M.W., Khalid, N.M., Zulkifli, R., Review on recent advances in sustainable additive manufacturing of aerospace structures. *Int. J. Precis. Eng. Manuf. Green Technol.*, 8, 3, 675–698, 2021, doi: 10.1007/s40684-021-00259-4.
32. Wang, P., Xu, X., Wang, J., Li, Y., Dai, K., Additive manufacturing for sustainable development: A state-of-the-art review. *J. Clean. Prod.*, 211, 707–723, 2019, doi: 10.1016/j.jclepro.2018.11.015.
33. Wang, X., Gao, X., Han, Y., A review on the development of additive manufacturing in sustainability. *J. Clean. Prod.*, 304, 127034, 2021, doi: 10.1016/j.jclepro.2021.127034.
34. Gao, X., Chen, H., Guo, W., Augmented reality for sustainable product design: A review. *J. Clean. Prod.*, 292, 126068, 2021, doi: 10.1016/j.jclepro.2021.126068.
35. Gomes, G., Carvalho, M.A., Pires, L.E., Assembly work instruction based on augmented reality: A literature review. *Int. J. Prod. Res.*, 58, 16, 5097–5118, 2020, doi: 10.1080/00207543.2019.1683574.
36. Li, K., Ma, Y., Chen, J., Zhang, J., Application of augmented reality in industrial energy conservation and emissions reduction. *Renew. Sustain. Energy Rev.*, 144, 111061, 2021, doi: 10.1016/j.rser.2021.111061.
37. Rizvi, S.A., Sarker, S., Taha, M.Z., Paul, S.K., An augmented reality-based remote maintenance approach for sustainable manufacturing. *Int. J. Prod. Res.*, 1–20, 2021, doi: 10.1080/00207543.2021.1930498.

38. Ghisellini, P., Cialani, C., Ulgiati, S., A review on circular economy: The expected transition to a balanced interplay of environmental and economic systems. *J. Clean. Prod.*, 114, 11–32, 2016, doi: 10.1016/j.jclepro.2015.09.007.
39. Ozcan, O.E. and Sivrikaya, F., Green manufacturing strategies for sustainable future. *J. Clean. Prod.*, 314, 128195, 2021, doi: 10.1016/j.jclepro.2021.128195.
40. Shishoo, R., *Waste minimization and clean technology in chemical process industries*, CRC Press, Boca Raton, 2018.
41. Tornese, F., Borsato, M., Colla, V., D'Amato, R., Life cycle assessment (LCA) as a tool to support sustainable manufacturing: A review of the state of the art. *Sustainability*, 11, 9, 2576, 2019, doi: 10.3390/su11092576.
42. European Commission, Green manufacturing, 2014, Retrieved from https://ec.europa.eu/growth/sectors/raw-materials/industries/green-manufacturing_en.
43. UNIDO, Green manufacturing, 2019, Retrieved from https://www.unido.org/our-focus/advancing-economic-competitiveness/sustainable-production-and-consumption/green-manufacturing.
44. Zailani, S.H.M., Sohail, M.S., Jayaraman, K., Green manufacturing practices and performance: A case of Malaysian automotive industry. *J. Clean. Prod.*, 204, 13–21, 2018, doi: 10.1016/j.jclepro.2018.08.201.
45. Geng, Y., Sarkis, J., Ulgiati, S., Toward a national circular economy indicator system in China: An evaluation and critical analysis. *J. Ind. Ecol.*, 14, 2, 314–326, 2010, doi: 10.1111/j.1530-9290.2009.00203.x.
46. Hervani, A.A., Helms, M.M., Sarkis, J., Performance measurement for green supply chain management. *Benchmarking Int. J.*, 12, 4, 330–353, 2005, doi: 10.1108/14635770510612721.
47. Pagell, M. and Wu, Z., Building a more complete theory of sustainable supply chain management using case studies of 10 exemplars. *J. Supply Chain Manage.*, 45, 2, 37–56, 2009, doi: 10.1111/j.1745-493x.2009.03169.x.
48. Sarkis, J., Zhu, Q., Lai, K.-H., An organizational theoretic review of green supply chain management literature. *Int. J. Prod. Econ.*, 130, 1, 1–15, 2010, doi: 10.1016/j.ijpe.2010.01.010.
49. Singh, R.K., Murty, H.R., Gupta, S.K., Dikshit, A.K., An overview of sustainability assessment methodologies. *Ecol. Indic.*, 69, 252–267, 2016, doi: 10.1016/j.ecolind.2016.04.043.
50. Vachon, S. and Klassen, R.D., Extending green practices across the supply chain: The impact of upstream and downstream integration. *Int. J. Oper. Prod. Manage.*, 26, 7, 795–821, 2006, doi: 10.1108/014435706106722.
51. Li, Y., Cao, Y., Li, J., An IoT-Based Energy Management System for Green Manufacturing. *IEEE Internet Things J.*, 8, 2, 1529–1541, 2021.
52. Zhang, X., Wang, Y., Lu, Y., A Review of Artificial Intelligence in Green Manufacturing: Practice, Applications, and Opportunities. *J. Clean. Prod.*, 258, 120771, 2020.

53. Liu, J., Zhao, Y., Cai, Y., Liu, Y., Digital Twin-Based Manufacturing Green Performance Evaluation: A Method and Case Study. *J. Clean. Prod.*, 283, 125425, 2021.
54. Liu, X., Wang, Y., Zhou, K., Bai, Y., Additive Manufacturing for Sustainable Development: A Comprehensive Review. *J. Clean. Prod.*, 252, 119826, 2020.
55. Zhang, Y., Li, J., Zhang, X., Li, Y., An Augmented Reality-Based Manufacturing Training System for Reducing Human Errors. *Comput. Ind.*, 126, 103466, 2021.
56. Jagtap, S., Rahimifard, S., Nabhani, F., Digital lean manufacturing: a new paradigm for Industry 4.0. *Procedia Manuf.*, 25, 266–271, 2018.
57. Linton, J.D., Klassen, R., Jayaraman, V., Sustainable supply chains: an introduction. *J. Oper. Manage.*, 25, 6, 1075–1082, 2007.
58. Schulte, R.D., Pereira, R.M., Anand, S., Verma, R., Bandyopadhyay, T., Combining Industry 4.0 and Sustainable Manufacturing Practices for Reducing Energy Consumption and Waste Generation: An Empirical Study. *Sustainability*, 13, 3, 1043, 2021.
59. World Economic Forum, The future of manufacturing: making things in a changing world, 2018, Retrieved from https://www.weforum.org/reports/the-future-of-manufacturing-making-things-in-a-changing-world.
60. Feng, S., Liu, X., Yang, S., Luo, X., Green and digital supply chain management: A review and future directions. *Int. J. Prod. Res.*, 58, 7, 2111–2131, 2020.
61. Gebauer, H., Worch, H., Truffer, B., The impact of digitalization and automation on employment in manufacturing industries. *Int. J. Oper. Prod. Manage.*, 40, 1, 23–43, 2020.
62. Gunasekaran, A., Papadopoulos, T., Dubey, R., Wamba, S.F., Childe, S.J., Hazen, B., Akter, S., Big data and predictive analytics for supply chain and organizational performance. *J. Bus. Res.*, 99, 325–338, 2019.
63. Müller, J.M., Buliga, O., Voigt, K., II, Fortune favors the prepared: How SMEs approach business model innovations in Industry 4.0. *Technol. Forecast. Soc. Change*, 132, 2–17, 2018.
64. Zhu, Q., Sarkis, J., Lai, K.H., Green supply chain management and Industry 4.0: An integrated framework and future research directions. *Int. J. Prod. Res.*, 57, 15-16, 4714–4733, 2019.
65. Ellen MacArthur Foundation, What is the circular economy?, 2019, Retrieved from https://www.ellenmacarthurfoundation.org/circular-economy/what-is-the-circular-economy.
66. Wang, F., Gao, R.X., Li, X., Digital twin-enabled sustainable manufacturing: A review. *J. Clean. Prod.*, 213, 1–15, 2019.
67. Bhattacharyya, S., Chattopadhyay, R., Ashour, A.S., 5G and sustainable manufacturing: A comprehensive review. *J. Clean. Prod.*, 256, 120340, 2020.
68. Bikas, H., Stavropoulos, P., Chryssolouris, G., Additive manufacturing methods and modeling approaches: A critical review. *Int. J. Precis. Eng. Manuf. Green Technol.*, 5, 2, 209–221, 2018.

69. Gao, R.X., Xu, X., Li, X., Wang, L., AI-enabled sustainable manufacturing: Recent progress and future trends. *J. Manuf. Syst.*, 56, 70–80, 2020.
70. Hu, H., Huang, J., Guo, L., Zhu, Z., Application of Augmented Reality Technology in Intelligent Manufacturing. *Int. J. Adv. Manuf. Technol.*, 94, 5-8, 1815–1825, 2018.
71. Li, X., Xu, X., Zhao, Y., Chen, B., Li, L., Digital Twins for Sustainable Manufacturing Systems: A Framework. *J. Clean. Prod.*, 172, 4235–4248, 2018.
72. Sharma, A., Weerakkody, V., El-Haddadeh, R., Blockchain Technology for Smart and Sustainable Supply Chain Management: A Systematic Literature Review and Future Research Directions. *Int. J. Inf. Manage.*, 49, 82–99, 2019.
73. Wang, Q., Pan, Y., Chen, B., Yu, Y., Huang, G.Q., Internet of Things-Enabled Intelligent Energy Management for Sustainable Manufacturing. *J. Clean. Prod.*, 220, 121–133, 2019.
74. Geissdoerfer, M., Savaget, P., Bocken, N.M., Hultink, E.J., The Circular Economy – A new sustainability paradigm? *J. Clean. Prod.*, 143, 757–768, 2017.
75. Hilty, L.M., Aebischer, B., Andersson-Sköld, Y., Bjarlert, A.L., Hult, M., Jägerbrand, A.K., Rueda, X., Product life extension: technological, environmental and social challenges and opportunities. *GAIA-Ecological Perspect. Sci. Soc.*, 28, 2, 108–114, 2019.
76. Jia, W., Li, D., Li, X., Li, W., Li, Z., Cui, Y., Big data and artificial intelligence for green manufacturing: challenges and opportunities. *J. Clean. Prod.*, 244, 118752, 2020.
77. Liu, X., Wang, Z., Shi, S., Zhang, Q., Green logistics in sustainable supply chain management: A systematic literature review and future research directions. *J. Clean. Prod.*, 258, 120511, 2020.
78. Mishra, R., Singh, R.K., Kumar, A., Kumar, P., Sustainable materials for green manufacturing: A review. *Mater. Today Proc.*, 21, 400–408, 2020.
79. Chisti, Y., Biomanufacturing - Opportunities and challenges. *Appl. Microbiol. Biotechnol.*, 102, 19, 7915–7925, 2018.
80. International Energy Agency, Renewables 2020: Analysis and forecast to 2025, IEA, Istanbul, Turkey, 2021.
81. Kirchherr, J., Reike, D., Hekkert, M., Resources, conservation & recycling. *Conceptualizing Circular Economy An Anal. 114 Def.*, 127, 221–232, 2018.
82. Wohlers Associates, Wohlers Report 2020: 3D printing and additive manufacturing state of the industry, Wohlers Associates, Inc, Fort Collins, Colorado, USA, 2020.
83. Steen, J. and Seepersad, C.C., A Review of Robotics and Automation in Green Manufacturing: Challenges, Opportunities, and Potential Solutions. *J. Clean. Prod.*, 252, 119723, 2020.

10

Smart Manufacturing for a Sustainable Future: A Review

Maninder Singh, Mukhtiar Singh, Jasvinder Singh* and Harjit Singh

School of Mechanical Engineering, Lovely Professional University, Punjab, India

Abstract

Smart manufacturing has emerged as a promising approach to optimize manufacturing processes, reduce waste, and minimize energy consumption for a sustainable future. In this abstract, we highlight the potential benefits of smart manufacturing for a green future. We discuss recent research and case studies on energy-efficient manufacturing processes, green supply chain management, waste reduction, renewable energy integration, green product design, circular economy, water conservation, and green data centers. The integration of advanced technologies such as artificial intelligence, the Internet of Things (IOT), and robotics can help reduce environmental impact and promote sustainability. Smart manufacturing can enable the transition to a circular economy where resources are reused, recycled, or repurposed, minimizing waste and promoting sustainability. Future directions of smart manufacturing for a sustainable future will require addressing challenges related to the adoption of new technologies, data management, and workforce development. Overall, smart manufacturing can play a significant role in promoting a green future by optimizing processes, reducing waste, and promoting sustainability.

***Keywords*:** Smart manufacturing, green supply chain management, renewable energy, waste conservation

**Corresponding author*: jasvinder.15854@lpu.co.in

Chandan Deep Singh, Talwinder Singh and Davinder Singh (eds.) Digitization and Manufacturing Performance: An Environmental Perspective, (265–280) © 2025 Scrivener Publishing LLC

10.1 Introduction

Smart manufacturing has the potential to revolutionize the manufacturing industry by improving efficiency, reducing waste, and promoting sustainability. Smart manufacturing is an approach that utilizes advanced technologies, such as artificial intelligence, the Internet of Things (IoT), and robotics to optimize manufacturing processes. In this paper, recent research and case studies in the field of smart manufacturing for a green future is reviewed. Also, the challenges and future directions of smart manufacturing for a sustainable future are discussed.

10.1.1 Smart Manufacturing for a Green Future

Energy-efficient manufacturing processes: Smart manufacturing can help reduce the energy consumption of manufacturing processes through the use of advanced technologies such as machine learning and IOT sensors. For instance, a study conducted by Geffen *et al.* [1] showed that the use of machine learning algorithms can reduce energy consumption in machining processes by up to 30%.

10.1.2 Green Supply Chain Management

Smart manufacturing can enable companies to monitor their supply chain and ensure that they are using environmentally friendly materials and processes. For example, a study conducted by Carter and Rogers [3] demonstrated that the use of IoT sensors can help monitor the environmental performance of suppliers and promote sustainable practices.

10.1.3 Waste Reduction

Smart manufacturing can help reduce waste by optimizing production processes and reducing the number of raw materials needed. For instance, a study conducted by Roy *et al.* [2] showed that the use of a digital twin simulation model can reduce the waste generated in manufacturing processes by up to 20%.

10.1.4 Renewable Energy Integration

Smart manufacturing can help integrate renewable energy sources into the manufacturing process, such as solar or wind power. For example, a study conducted by Ang *et al.* [5] demonstrated that the use of solar power in manufacturing processes can reduce greenhouse gas emissions by up to 50%.

10.1.5 Green Product Design

Smart manufacturing can help companies design and produce environmentally friendly products that are sustainable and energy-efficient. For instance, a study conducted by Bird *et al.* [6] showed that the use of additive manufacturing can reduce the environmental impact of manufacturing processes by up to 90%.

10.1.6 Circular Economy

Smart manufacturing can enable the transition to a circular economy where resources are reused, recycled, or repurposed, minimizing waste and promoting sustainability. For example, a study conducted by Sharma *et al.* [7] demonstrated that the use of a closed-loop manufacturing system can reduce waste generation and promote sustainability.

10.1.7 Water Conservation

Smart manufacturing can help reduce water consumption in the manufacturing process by optimizing water usage and recycling water. For instance, a study conducted by Krishnan *et al.* [8] showed that the use of IoT sensors and machine learning algorithms can reduce water consumption in the manufacturing process by up to 40%.

Green data centers: Smart manufacturing can help reduce the carbon footprint of data centers through the use of renewable energy sources and energy-efficient technologies. For example, a study conducted by Mondal *et al.* [9] showed that the use of renewable energy sources such as wind power can significantly reduce the carbon footprint of data centers.

10.2 Smart Manufacturing for Green Future

Energy-efficient manufacturing processes have become an important area of research and development in recent years due to their potential to reduce energy consumption, cut costs, and improves sustainability in the manufacturing industry. In this review paper, we will explore the various methods and techniques used in energy-efficient manufacturing processes and their potential impact on the industry.

10.2.1 Energy-Efficient Equipment and Machinery

Using energy-efficient equipment and machinery is one of the most effective ways to reduce energy consumption in manufacturing processes. Energy-efficient equipment can be designed to use less energy, operate at lower temperatures, and consume fewer resources. Examples of energy-efficient equipment include high-efficiency motors, variable frequency drives, and advanced control systems.

A study by the International Energy Agency (IEA) found that energy-efficient equipment can reduce energy consumption in manufacturing processes by up to 30% [10]. For example, replacing a standard motor with a high-efficiency motor can reduce energy consumption by up to 20%, while replacing traditional lighting with LED lighting can reduce energy consumption by up to 70% [1].

10.2.2 Process Optimization

Process optimization involves identifying and eliminating inefficiencies in manufacturing processes to reduce energy waste and improve productivity. Process optimization techniques include lean manufacturing, Six Sigma, and Total Productive Maintenance (TPM).

A study by the Lawrence Berkeley National Laboratory found that process optimization techniques can reduce energy consumption in manufacturing processes by up to 15% [1]. For example, optimizing production schedules to reduce idle time and minimize changeovers can reduce energy consumption by up to 10%, while optimizing equipment maintenance schedules can reduce energy consumption by up to 5%.

10.2.3 Energy Management Practices

Energy management practices involve monitoring, measuring, and controlling energy consumption in manufacturing processes to identify opportunities for improvement and implement energy-saving measures. Energy management practices include energy audits, energy benchmarking, and energy management systems (EMS).

A study by the U.S. Department of Energy found that implementing energy management practices can reduce energy consumption in manufacturing processes by up to 20% [29]. For example, conducting an energy audit can identify energy-saving opportunities, such as upgrading equipment or optimizing processes, while implementing an energy management system can provide a framework for continuous improvement in energy efficiency.

10.2.4 Renewable Energy Sources

Using renewable energy sources, such as solar, wind, or hydropower, can also help reduce energy consumption in manufacturing processes. Renewable energy can be used to power equipment and machinery or provide heat for manufacturing processes.

A study by the International Renewable Energy Agency (IRENA) found that using renewable energy sources in manufacturing processes can reduce energy consumption by up to 50% [15]. For example, installing solar panels on the roof of a manufacturing facility can provide a significant portion of the energy needed to power the facility, while using biomass for heat in manufacturing processes can reduce reliance on fossil fuels.

10.3 Green Supply Chain Management

Green supply chain management (GSCM) is an approach to managing the environmental impact of supply chain activities. It involves integrating sustainability practices into every aspect of the supply chain, from product design to end-of-life disposal [4]. In this section, we will explore the key concepts of GSCM and the benefits it can bring to businesses and society.

One of the key concepts of GSCM is sustainable product design. This involves designing products that minimize their environmental impact throughout their lifecycle, from production to disposal [3]. This includes using sustainable materials, reducing waste, and ensuring that products can be easily disassembled and recycled. Another important concept is green procurement, which involves selecting suppliers and materials that meet environmental standards and reduce the environmental impact of the supply chain [4]. This includes selecting suppliers that use renewable energy, minimize waste, and have environmentally friendly production processes.

Energy efficiency is another important aspect of GSCM. It involves reducing the energy consumption of supply chain activities, including manufacturing, transportation, and storage. This can be achieved through the use of energy-efficient equipment and processes, as well as the adoption of renewable energy sources [11]. Waste reduction is also an important aspect of GSCM, which involves minimizing waste throughout the supply chain, from raw material production to end-of-life disposal. This includes reducing packaging waste, implementing recycling programs, and reducing product returns [3].

GSCM can bring numerous benefits to businesses and society. For example, GSCM can lead to cost savings through reduced energy and resource consumption, waste reduction, and increased efficiency. It can also provide a competitive advantage by demonstrating a commitment to sustainability and meeting the increasing demand for sustainable products and services. Additionally, GSCM can improve relationships with stakeholders, including customers, suppliers, and regulatory bodies, by demonstrating a commitment to environmental responsibility. Furthermore, GSCM can significantly reduce the environmental impact of supply chain activities, including greenhouse gas emissions, water consumption, and waste generation [11].

Several case studies demonstrate the benefits of GSCM. For instance, Wal-Mart implemented GSCM practices, such as packaging reduction, energy efficiency, and waste reduction, which resulted in cost savings and improved relationships with suppliers and customers. Toyota also implemented GSCM practices, such as sustainable product design and energy efficiency, which resulted in cost savings and reduced environmental impact. Similarly, Patagonia implemented GSCM practices, such as sustainable product design and green procurement, which resulted in improved stakeholder relations and increased sales of sustainable products [4].

10.4 Waste Reduction

It is an important aspect of sustainable development that involves minimizing waste generation and promoting the efficient use of resources. In this section, we will explore the key concepts of waste reduction and the benefits it can bring to businesses and society.

One of the key concepts of waste reduction is the "waste hierarchy," which prioritizes waste management strategies in order of preference: reduce, reuse, recycle, and dispose [13]. The first step is to reduce waste generation by preventing waste from being created in the first place. This can be achieved through the use of environmentally friendly production processes, product design that minimizes waste, and the promotion of sustainable consumption patterns. The second step is to promote the reuse of products and materials by designing products that are durable and can be easily repaired or repurposed [14]. The third step is to recycle materials that cannot be reused by collecting and processing them into new products. The final step is to dispose of any remaining waste in a safe and environmentally responsible manner.

Another important concept of waste reduction is the "circular economy," which involves designing products and processes that promote the efficient use of resources and minimize waste generation. This includes using renewable materials, designing products for disassembly and recycling, and promoting a closed-loop system of material use [20].

Waste reduction can bring numerous benefits to businesses and society. For example, waste reduction can lead to cost savings through reduced waste disposal fees, decreased raw material and energy consumption, and increased efficiency. It can also provide a competitive advantage by demonstrating a commitment to sustainability and meeting the increasing demand for sustainable products and services [12]. Additionally, waste reduction can significantly reduce the environmental impact of waste generation, including greenhouse gas emissions, water consumption, and land use [13].

Several case studies demonstrate the benefits of waste reduction. For example, Toyota implemented waste reduction practices, such as using reusable containers and reducing packaging waste, which resulted in cost savings and improved efficiency. Dow Chemical implemented waste reduction practices, such as using renewable materials and designing products for disassembly, which resulted in cost savings and increased customer satisfaction. Similarly, Interface implemented waste reduction practices, such as using recycled materials and designing products for disassembly, which resulted in improved stakeholder relations and increased sales of sustainable products [20].

10.5 Renewable Energy Integration

It refers to the process of integrating renewable energy sources, such as wind and solar power, into the existing energy infrastructure. This integration can present significant challenges, as renewable energy sources are often intermittent and decentralized, and require advanced control systems to manage their integration into the grid. In this section, we will explore the key concepts of renewable energy integration and the challenges and opportunities it presents, with references and citations.

One of the key challenges of renewable energy integration is the intermittent nature of renewable energy sources. Solar and wind power generation is dependent on weather conditions, and their output can fluctuate rapidly. This creates challenges for grid operators, who need to balance the supply and demand of electricity in real-time to maintain system stability. To address this challenge, advanced control systems, such as energy storage, demand response, and advanced forecasting, can be used to manage the variability of renewable energy sources [16].

Another challenge of renewable energy integration is the need for a decentralized energy infrastructure. Renewable energy sources are often located far from population centers, and their integration into the grid requires the development of new transmission infrastructure. This can be expensive and time-consuming, and requires collaboration between grid operators, regulators, and the renewable energy industry [15].

Despite these challenges, renewable energy integration presents significant opportunities for the energy sector. Renewable energy sources can reduce greenhouse gas emissions, improve energy security, and create new economic opportunities. Additionally, renewable energy sources can improve energy access in remote and underserved areas, providing a pathway for sustainable development [15].

Several countries have made significant progress in renewable energy integration. For example, Denmark has successfully integrated a high proportion of wind power into its energy system, reaching a peak of 57% wind power in 2020 [15]. Germany has also made significant progress in renewable energy integration, reaching a peak of 46% renewable electricity in 2020 [16]. China has also made significant investments in renewable energy integration, with a target of reaching 50% renewable electricity by 2050.

10.6 Green Product Design

It refers to the process of designing products that minimize negative environmental impacts throughout their entire lifecycle, from raw material extraction to disposal. In this section, we will explore the key concepts of green product design and the benefits it provides, with references and citations.

Green product design considers environmental impact at all stages of the product lifecycle. The design process involves considering factors such as material selection, energy efficiency, waste reduction, and end-of-life disposal. By taking these factors into account, green product design can reduce environmental impact, conserve natural resources, and improve product performance [18].

One key aspect of green product design is the use of sustainable materials. This includes selecting materials that have a low environmental impact, such as renewable resources or recycled materials, and minimizing waste throughout the manufacturing process. In addition, green product design considers the end-of-life of the product, including the potential for recycling or reuses [17].

Another key aspect of green product design is energy efficiency. This includes designing products that use less energy during their use phase, as well as reducing energy consumption during the manufacturing process. By improving energy efficiency, green product design can reduce greenhouse gas emissions and conserve natural resources [18].

Green product design also considers the social and ethical implications of product design. This includes issues such as worker health and safety, human rights, and fair labor practices. By designing products that consider these issues, companies can improve their social and environmental impact and build consumer trust [17].

Green product design provides numerous benefits for companies and consumers. For companies, green product design can reduce costs, improve brand reputation, and provide a competitive advantage in the marketplace. For consumers, green products offer improved performance, reduced environmental impact, and healthier living environments [17].

Several companies have successfully implemented green product design. For example, Patagonia, a clothing company, has implemented a program to reduce the environmental impact of its products. The company has implemented a system to track the environmental impact of its products throughout their lifecycle, and uses this information to make design decisions that minimize environmental impact [19].

10.7 Circular Economy

It is an economic model that aims to minimize waste and promote the efficient use of resources by keeping materials and products in use for as long as possible. This section will provide detailed information on circular economy and its benefits, with references and citations.

The circular economy model is based on the principles of reduce, reuse, and recycle. This model emphasizes the importance of designing products and materials with the intention of keeping them in use for as long as possible, minimizing waste, and promoting the efficient use of resources [20].

In a circular economy, waste is seen as a valuable resource that can be reused, recycled, or repurposed. This model aims to close the loop on resource use, eliminating the need for disposal and reducing the environmental impact of production and consumption [22].

The benefits of a circular economy are numerous. By reducing waste and promoting the efficient use of resources, a circular economy can reduce greenhouse gas emissions, conserve natural resources, and promote sustainable economic growth [20].

Circular economy has gained traction globally, with several countries implementing policies and initiatives to promote circular economy practices. The European Union, for example, has developed a circular economy action plan, which outlines strategies to promote the circular economy, including waste reduction, recycling, and eco-design [21].

Several companies have also embraced the circular economy model. For example, Philips, a consumer electronics company, has implemented a circular economy program to promote the reuse and recycling of its products. The company has developed a range of products designed for disassembly and reuse, and has implemented a take-back program to collect and recycle its products at the end of their life [23].

10.8 Water Conservation

It is the practice of using water efficiently to reduce waste and ensure its availability for future generations. This section will provide detailed information on water conservation and its benefits, with references and citations.

Water conservation is becoming increasingly important due to the growing global population and the increasing demand for water in various sectors, such as agriculture, industry, and households [24]. By conserving water, we can reduce the strain on water resources, protect aquatic ecosystems, and reduce the energy required for water treatment and transportation [26].

There are several ways to conserve water, including:

- Fixing leaks and repairing faulty pipes and faucets
- Installing water-efficient fixtures and appliances
- Implementing water-saving practices, such as taking shorter showers and turning off the tap when brushing teeth
- Using drought-tolerant plants in landscaping
- Collecting and using rainwater for irrigation and other non-potable uses

Water conservation practices can benefit both the environment and the economy. For example, water-efficient practices can reduce the need for costly water infrastructure and treatment, and can improve the reliability of water supplies for agricultural and industrial users [29].

Many countries and organizations have developed policies and initiatives to promote water conservation. The United Nations, for example, has developed the Sustainable Development Goal 6, which aims to ensure the availability and sustainable management of water and sanitation for all [25].

In the United States, the Environmental Protection Agency (EPA) has developed the WaterSense program, which promotes water-efficient products and practices in homes, businesses, and industries [26].

10.9 Green Data Centers

These are facilities that are designed and operated with energy efficiency and environmental sustainability in mind. This section will provide detailed information on green data centers, their benefits, and some of the strategies that can be used to make them more environmentally friendly, with references and citations.

Data centers are facilities that house computer systems, servers, and other equipment that support the processing and storage of digital data. As the demand for digital services and data storage continues to grow, the energy consumption of data centers has become a major concern for environmental sustainability. Green data centers aim to address this issue by optimizing energy efficiency and reducing the environmental impact of data center operations.

There are several strategies that can be used to make data centers more environmentally friendly, including:

- Using energy-efficient equipment and cooling systems
- Maximizing server utilization through virtualization and consolidation
- Implementing renewable energy sources, such as solar and wind power
- Recycling and reusing electronic waste
- Using advanced monitoring and management systems to optimize energy efficiency

The benefits of green data centers are numerous. They can help reduce greenhouse gas emissions, lower energy consumption, and decrease the carbon footprint of data center operations. In addition, green data centers can help companies save money on energy costs and improve their reputation by demonstrating their commitment to environmental sustainability.

Many companies and organizations have implemented green data center strategies. For example, Google has made significant investments in renewable energy sources and has implemented advanced cooling systems to improve energy efficiency in its data centers [27]. Microsoft has also made significant strides in energy efficiency and sustainability, with a goal of achieving carbon neutrality by 2030 [28].

10.10 Conclusion

Energy-efficient manufacturing processes are essential for reducing energy consumption, cutting costs, and improving sustainability in the manufacturing industry. By using energy-efficient equipment and machinery, optimizing processes, implementing energy management practices, and using renewable energy sources, manufacturers can significantly reduce their environmental impact and contribute to global efforts to mitigate climate change.

GSCM is an important approach to managing the environmental impact of supply chain activities. By integrating sustainability practices into every aspect of the supply chain, businesses can achieve cost savings, gain a competitive advantage, and reduce their environmental impact. As the demand for sustainable products and services continues to grow, GSCM will become increasingly important for businesses looking to meet the expectations of their customers, stakeholders, and society.

Waste reduction is an important aspect of sustainable development that involves minimizing waste generation and promoting the efficient use of resources. By implementing waste reduction practices, businesses can achieve cost savings, gain a competitive advantage, and reduce their environmental impact. As the demand for sustainable products and services continues to grow, waste reduction will become increasingly important for businesses looking to meet the expectations of their customers, stakeholders, and society.

Renewable energy integration presents significant challenges and opportunities for the energy sector. While the integration of renewable energy sources into the existing energy infrastructure requires advanced control systems and new transmission infrastructure, the benefits of renewable energy integration, such as reducing greenhouse gas emissions, improving energy security, and creating new economic opportunities, are significant. As the demand for renewable energy continues to grow, renewable energy integration will become increasingly important for the energy sector to meet the expectations of customers, regulators, and society.

Green product design is a critical component of sustainable manufacturing. By considering environmental impact, energy efficiency, and social and ethical issues, companies can design products that reduce environmental impact, improve performance, and build consumer trust. As consumer demand for sustainable products continues to grow, green product design will become increasingly important for companies looking to remain competitive in the marketplace.

The circular economy is an economic model that promotes the efficient use of resources and minimizes waste. By embracing the principles of reduce, reuse, and recycle, companies and governments can promote sustainable economic growth, reduce environmental impact, and conserve natural resources.

Water conservation is a critical practice that can help protect our water resources, reduce waste, and promote sustainable economic growth. By implementing water-efficient practices and technologies, individuals, businesses, and governments can conserve water and ensure its availability for future generations.

Green data centers are an important strategy for promoting energy efficiency and environmental sustainability in the digital age. By implementing energy-efficient technologies and practices, companies and organizations can reduce their environmental impact and improve their bottom line.

References

1. Geffen, C., Price, L., Schiller, S., Energy Efficiency Improvement and Cost Saving Opportunities for the U.S, in: *Iron and Steel Industry: An ENERGY STAR® Guide for Energy and Plant Managers*, Lawrence Berkeley National Laboratory, Berkeley, CA, 2016.
2. Roy, R.B., Mishra, D., Pal, S.K., Chakravarty, T., Panda, S., Chandra, M.G., Misra, S., Digital twin: current scenario and a case study on a manufacturing process. *Int. J. Adv. Manuf. Technol.*, 107, 3691–3714, 2020.

3. Carter, C.R. and Rogers, D.S., A framework of sustainable supply chain management: moving toward new theory. *Int. J. Phys. Distrib. Logist. Manage.*, 38, 5, 360–387, 2008.
4. Sarkis, J., A boundaries and flows perspective of green supply chain management. *Supply Chain Manage. An Int. J.*, 17, 2, 202–216, 2012.
5. Ang, T.Z., Salem, M., Kamarol, M., Das, H.S., Nazari, M.A., Prabaharan, N., A comprehensive study of renewable energy sources: Classifications, challenges and suggestions. *Energy Strategy Rev.*, 43, 100939, 2022.
6. Bird, L., Milligan, M., Lew, D., Integrating variable renewable energy: Challenges and solutions (No. NREL/TP-6A20-60451). National Renewable Energy Lab. NREL, Golden, CO United States, 2013.
7. Sharma, M., Vadalkar, S., Antony, R., Chavan, G., Tsagarakis, K.P., Can Industry 4.0-enabled smart manufacturing help firms in emerging economies move toward carbon-neutrality? *Comput. Ind. Eng.*, 192, 110238, 2024.
8. Krishnan, S.R., Nallakaruppan, M.K., Chengoden, R., Koppu, S., Iyapparaja, M., Sadhasivam, J., Sethuraman, S., Smart water resource management using Artificial Intelligence—A review. *Sustainability*, 14, 20, 13384, 2022.
9. Mondal, S., Faruk, F.B., Rajbongshi, D., Efaz, M.M.K., Islam, M.M., GEECO: Green Data Centers for Energy Optimization and Carbon Footprint Reduction. *Sustainability*, 15, 21, 15249, 2023.
10. Canton, H., International energy agency—IEA, in: *The Europa Directory of International Organizations*, pp. 684–686, Routledge, London, United Kingdom, 2021.
11. Walker, H. and Jones, N., Sustainable supply chain management across the UK private sector. *Supply Chain Manage. An Int. J.*, 17, 1, 15–28, 2012.
12. Ellen MacArthur Foundation, The circular economy, 2022, Retrieved from https://www.ellenmacarthurfoundation.org/circular-economy/what-is-the-circular-economy.
13. Environmental Protection Agency (EPA), Waste management hierarchy, 2022, Retrieved from https://www.epa.gov/waste/waste-management-hierarchy.
14. Industrial Waste Reduction Program (IWRP), Waste reduction case studies, 2022, Retrieved from https://www.iwrp.org/waste-reduction-case-studies/.
15. International Renewable Energy Agency (IRENA), Renewable Energy Integration: Challenges and Solutions, 2019, Retrieved from https://www.irena.org/-/media/Files/IRENA/Agency/Publication/2019/Mar/IRENA_REmap_REintegration_2019.pdf.
16. International Renewable Energy Agency (IRENA), Renewable Energy Statistics 2021, 2021, Retrieved from https://www.irena.org/-/media/Files/IRENA/Agency/Publication/2021/Mar/IRENA_Renewable_Energy_Statistics_2021.pdf.
17. Bocken, N., Allwood, J.M., Willett, I., Evans, S., A review of design for sustainability (DfS) approaches and their implementation in the context of sustainable development. *J. Clean. Prod.*, 132, 76–84, 2016.

18. International Organization for Standardization (ISO), ISO 14006:2011 - Environmental management systems - Guidelines for incorporating ecodesign, 2019, Retrieved from https://www.iso.org/standard/52911.html.
19. Patagonia, Footprint Chronicles®, n.d, Retrieved from https://www.patagonia.com/footprint-chronicles/.
20. Ellen MacArthur Foundation, Towards the Circular Economy: Economic and business rationale for an accelerated transition, 2013, Retrieved from https://www.ellenmacarthurfoundation.org/publications/towards-the-circular-economy-volume-1-an-economic-and-business-rationale-for-an-accelerated-transition.
21. European Commission, Circular Economy Action Plan, 2020, Retrieved from https://ec.europa.eu/environment/circular-economy/index_en.htm.
22. Kirchherr, J., Reike, D., Hekkert, M., Conceptualizing the circular economy: An analysis of 114 definitions. *Resour. Conserv. Recycl.*, 127, 221–232, 2017.
23. Philips, Circular Economy, 2019, Retrieved from https://www.philips.com/a-w/about/sustainability/circular-economy.html.
24. Environmental Protection Agency (EPA), WaterSense, 2022, Retrieved from https://www.epa.gov/watersense.
25. UN-Water, Sustainable Development Goal 6, n.d, 2017, Retrieved from https://www.unwater.org/water-facts/sustainable-development-goal-6/.
26. WaterSense, Why Water Efficiency?, n.d, 2017, Retrieved from https://www.epa.gov/watersense/why-water-efficiency.
27. Google, Sustainability at Google, 2022, Retrieved from https://sustainability.google/.
28. Microsoft, Microsoft Sustainability, 2022, Retrieved from https://www.microsoft.com/en-us/corporate-responsibility/sustainability.
29. U.S. Department of Energy (DOE), Energy Efficient Data Centers, 2022, Retrieved from https://www.energy.gov/eere/femp/energy-efficient-data-centers.

11

Smart Manufacturing for a Greener Future

Harpreet Sharma[1]*, Chandan Deep Singh[2] and Kanwaljeet Singh[3]

[1]*Yadavindra Department of Engineering, Guru Kashi Campus, Talwandi Sabo, Punjab, India*
[2]*Punjabi University, Department of Mechanical Engineering, Patiala, India*
[3]*Punjab State Aeronautical Engineering Colleges, Patiala Civil Aerodrome, Patiala, India*

Abstract
Smart manufacturing is an advanced approach that utilizes digital technologies, automation, and data analytics to optimize manufacturing processes and enhance productivity. Figure 11.1 represents a schematic of smart manufacturing components. This chapter explores the concept of smart manufacturing and its potential to contribute to a greener future. It highlights the role of smart manufacturing in improving resource efficiency, reducing waste, and minimizing environmental impact. The chapter also discusses the integration of renewable energy sources and the implementation of sustainable practices in smart manufacturing systems.

Keywords: Smart factory, industrial IoT (IIoT), energy optimization, predictive maintenance, digital twin, sustainable automation, green supply chain, carbon emissions tracking

11.1 Introduction and Meaning

Global climate change has grown into one of the major concerns in global industrialization and economic development. Carbon dioxide (CO2), one of the six Kyoto Protocol greenhouse gases (GHGs), is the most difficult to contain due to its closeness with energy generation and consumption. With an earlier awareness of climate change, developed countries tended to displace the resource and energy ntensive industries to other countries,

**Corresponding author*: harpreetsharma3211@gmail.com

Chandan Deep Singh, Talwinder Singh and Davinder Singh (eds.) Digitization and Manufacturing Performance: An Environmental Perspective, (281–298) © 2025 Scrivener Publishing LLC

thus emissions have not been reduced but relocated, which can be further magnified if production shifts to countries using carbon-intensive energy or low-energy-efficient technologies. Developing countries are responsible for 63% of GHGs emissions and get hit the hardest, i.e., 78% of the cost of climate change, and predicted to rise to 87% by 2035 [1]. For example, people in tropical countries are more exposed to natural disasters, e.g., storms and extreme weather. Poorer living conditions make them fragile to disaster strikes. Common measures of controlling GHG emissions are pricing the emissions and adopting cleaner energy. Even if a price can be offered, it is unlikely to convince developing countries to instantly give up fossil fuel or to pursue nuclear energy while ignoring non-proliferation and nuclear waste. A more acceptable solution, at least in the near future, is circular economy or to reduce emissions without substantial changes in energy infrastructure using intelligent technologies [2].

Manufacturing, a core of the product provision system and a major source of GHG and other emissions, has substantial influences on both economic and sustainability issues. Years ago, before the topic of Industry 4.0 become the dominating theme of the Hanover Trade Fair, its motto had been to integrate intelligent technologies and green methods [3]. The objective is to capture and monitor manufacturing activities and shift them into a greener and more eco-efficient paradigm while holding competing costs. A greener paradigm may offer opportunities for cost reduction and eco-friendly development. Industry 5.0, a new industrial paradigm from the European Commission's consensus has been proposed to strategically plan the production and economic growths accounting for the planetary boundaries and societal goals. But exploiting these opportunities involves more than technological issues, which requires identifying priorities and finding ways to align them with existing infrastructures and technologies. Over the last decade, intelligent technologies have been growing rapidly for increasing production efficiency and flexibility. There are many ways to benefit manufacturing facilities and workplace practices with improved environmental performance. It requires the current manufacturing system to go beyond producing goods and/or services for profit and highlights the importance of long-term service to environmental and societal benefits, including human-centricity, sustainability and resilience [4].

Evolving smart manufacturing systems into a greener paradigm in Industry 5.0 era is both timely and critical.

The implications of intelligent technologies to green objectives are deemed controversial and the greenness assessment of intelligent technologies also remains to be addressed. In this chapter, we propose "Smart manufacturing systems for greener future" as integrated manufacturing paradigm that uses intelligent techniques as the enabler for green objectives.

Figure 11.1 A schematic of smart manufacturing components.

Various green and sustainable objectives are potentially accompanied by intelligent technologies, and the sustainability of intelligent systems should also be assessed and enhanced by green metrics.

The primary goal of smart manufacturing for greener future is to integrate intelligent technologies and green methods to achieve superior control, productivity and sustainability in manufacturing systems.

Smart manufacturing framework is designed accordingly to encompass a continuously evolving group of methods and materials, from techniques for clean energy generation to the design of green products, as shown in Figure 11.2. The adaptation of intelligent technologies helps to better reach sustainable objectives while remaining competitive and value-adding to advanced manufacturing sectors.

Our Smart manufacturing for green future addresses the term 'green' in several aspects:

- To deliver greener products by changing industrial processes and product design.
- To green the manufacturing process by reducing emissions and waste, minimizing natural resource use, and improving the energy and manufacturing efficiency.
- To assess the greenness of abovementioned intelligent techniques.

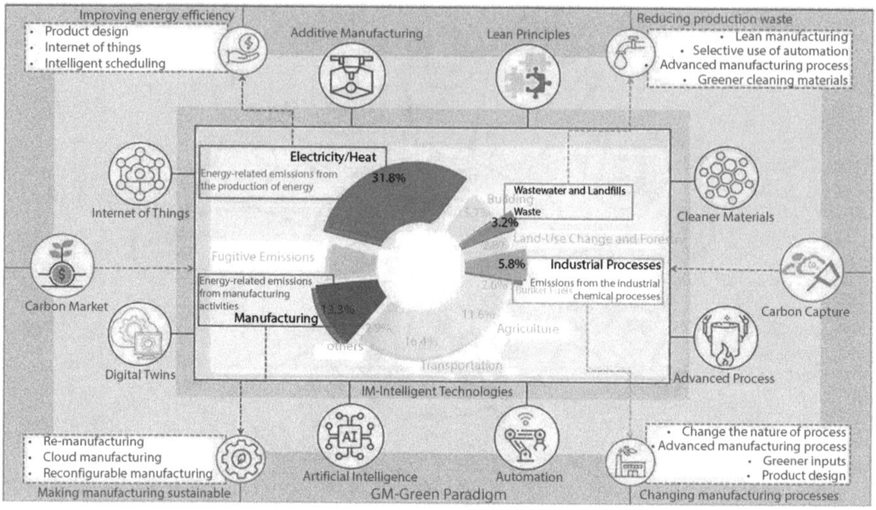

Figure 11.2 Framework of smart manufacturing for greener future.

11.2 Historical Background of Manufacturing

From the early age machining process to today's automated manufacturing industries, the manufacturing industries have achieved huge advancements. Going through various inventions and repetitive trials over the years, manufacturing and industrialization systems are now reached their fourth generation of the industrial revolution. During this period, there have been several improvements and changes as far. The timeline in Figure 11.3 shows the different ages of the industrial revolution highlighting major achievements/inventions which are identical to be the newer revolution.

While our society is promoting exponential technology growth, human activities and many resources (e.g., energy, materials, water) are increasingly resulting in climate change. The modern industry is facing two significant changes that will affect all perspectives of consumption and production of the 21st century, the fourth industrial revolution, and the implementation of the sustainable manufacturing approach. The fourth industrial revolution, also known as industry 4.0, is expected to transform not only the industrial production but also the whole society. Historically, society has experienced four major industrial transformations. This starts with the utilization of steam power and mechanical mechanisms in production in 1784, which represents the first industrial revolution. The second industrial revolution was realized when the concept of mass production and assembly lines was adopted in different industries, for example, Henry Ford, in 1870.

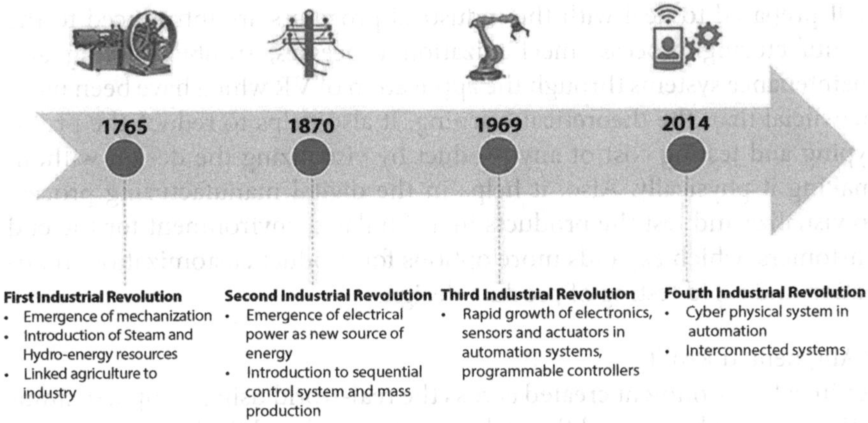

Figure 11.3 Industrial revolutions.

The third industrial revolution focused on automation and IT with the introduction of PLC and robotics in a variety of manufacturing stages in 1969. Currently, and since 2001, the industry is experiencing the fourth industrial revolution. This revolution is based on emergent technologies such as Big Data Analytics and the Internet of Things (IoT), enabling the development of an intelligent network that permeates all production stages and completely changes the relationship between companies and society.

11.2.1 Technologies Associated with Smart Manufacturing

Smart manufacturing integrates various technologies related to manufacturing, computing, virtualization, connectivity, data handling etc. The scope of smart manufacturing technologies has become broader due to the inter-operation of various technologies resulting in cost-effectiveness, time-saving, easy configuration, better understanding, quick response to market demand, flexibility and remote monitoring.

• Virtual Reality

Virtual Reality (VR) provides the facility for experiencing the computer-generated images and videos simulating into real-world activities. VR is a wearable device containing video device, audio device, positioning systems such as GPS, external connectivity to other devices and hardware to make the user experience his physical presence in the virtual environment created in simulation. VR has been used in the manufacturing systems for training the young engineers and technical graduates who are not well-prepared to deal with the industrial processes are introduced to the manufacturing process, mechanization processes, troubleshooting and maintenance systems through the application of VR which have been more beneficial than the theoretical learning. It also helps to reduce the prototyping and testing cost of any product by visualizing the design without making it physically. Also, it helps in the digital manufacturing process to visualize and test the products in a simulated environment for the end customers, which expands more options for product customization, renovation and rapid testing of product design.

- Augmented Reality

Artificial environment created across the real-world using computer simulation and can be realized through wearables and mobile devices. They use the technology of combining the real physical environment along with the computer-generated graphics to visualize artificially added components to the existing real-world scenario for training, simulation or validation purpose of some of the manufacturing designs before going for real manufacturing. This integration of simulated computer graphics to real-world scenarios helps to realize the product in an existing environment. Training to new employees and product testing with a demonstration of various conditions in the augmented environment has been found more efficient and time-saving [5].

- Cyber-Physical Systems

The industrial processes can be directly monitored and controlled through the use of the internet by the industrial manufacturing engineers, which enables the control engineer to access the control system of the industry from anywhere through cloud computing. Cyber-Physical System (CPS) collaborates the computational entities to the physical world and its ongoing processes utilizing data processing services available directly on the internet. The cloud-based SCADA system can also be defined as the CPS where the sensors, actuators and hardware devices fall under the physical devices and the software, communication system and information exchange through the internet falls under the cyber layer. The major application areas of CPS are implemented in aerospace, automotive, transportation and civil infrastructure [6].

- Additive Manufacturing

The Additive Manufacturing (AM) also known as 3D printing technology creates new horizons towards smart manufacturing technology. AM technology is flexible for customization, rapid prototyping, making spare parts quickly and on-site manufacturing which also saves huge time of production and cost in machine tools replacements and raw materials. AM also facilitates reverse engineering of any parts or product through 3D scanning and allows reconfiguration in design and quick reproduction for testing and validation is the major contribution of AM in smart manufacturing. AM technology is also being widely used in medical science for implants in dentistry and orthopedics for the replacement of injured body parts. In civil and architectural engineering, AM is utilized in prototyping and design testing of structures for cost-effectiveness and customer satisfaction [7].

- Big Data Analytics

The collection and analysis of data from various sources like production unit, enterprise, customer feedback and product request system, etc. help in real-time decision making for smart manufacturing. Today's manufacturers want their customers to share their feedback and personal view in the products they are using or intend to use, and from that information, manufacturers intend to focus their product design to address the wide range of customers. Big data analysis will benefit the manufacturer in the identification of current state and causes of product failures in real-time, driving the customers to buy their products by understanding their buying habits and requirements, and also, learn the potential of data-driven marketing for predictive manufacturing [8].

- Flexible and Reconfigurable Manufacturing Systems (FRMS)

Manufacturing system capable of adopting any variations occurred in built-in priority and functions due to the market demand changes or due to reconfiguration of product. They are focused on cost-effectiveness and quick response to rapid system changes. It is capable of production in small batch size and easily reconfigured for manufacture different variety of products which can be divided into two types; routine flexibility allows the manufacturing plant to produce new types of product in the same production line while machine flexibility allows to schedule the manufacturing routines of different machining stations, share the machining activity between them, replace the machining parts automatically. The deterministic characteristics of FRMS in smart manufacturing are modularity, integrability, flexibility, scalability and diagnosability [9].

- Artificial Intelligence

For better human-robot collaboration and reduce the human workforce in risk zones, to improve the maintenance system of the manufacturing system and detection of any failures in the machinery and the product, artificial intelligence is adapted in new generation manufacturing system. The artificial intelligence system is capable of self-decision, self-optimizing and automatic response to physical changes such as changing the production schedules, stopping or running of any machining units, machine tools automatic replacement and warning of any uncontrollable situations in time.

• IoT and IIoT
IoT is being used in common domestic applications like smart homes, transportation, logistics, healthcare, agriculture, human pets and vehicle tracking applications and IIoT is focused on the Industrial application of IoT connecting every physical entity with each other through the internet. In industries, the IIoT connects the physical entities like sensors, actuators and the entire process monitoring and control system into the internet cloud, enabling the interaction and cooperation of each entity together to reach the common goals. These interactions between each component help in production planning, predictive maintenance and faults localization, better human-machine interaction, intelligent process control for resource, tools and materials optimization. IIoT also allows the digital presentations of products, processes and factories to the customers for marketing and informative purposes too.

• Simulation
Simulation of plant operations to optimize the machine settings for preceding production line in a virtual environment without testing in physical world would save time and money in testing the production system. Simulations are used to mirror the physical world in a virtual model which can be used for production planning and scheduling according to obtained simulation results. Moreover, simulation has been implemented in SMS system these days which helped to analyze the design errors, production time, workforce and energy required for the entire production and facilitated the preparation of cost estimation, profit loss analysis, preparation of bill of quantity before proceeding into real manufacturing process.

11.3 Characteristics and Challenges of Smart Manufacturing Systems

The smart manufacturing systems are capable to cope up with various challenges and complexities faced by the existing industries, but there still exists some of the challenges during the implementation of smart manufacturing systems. Based on various dependent variables, smart manufacturing systems are considered to be faced by security issues, lack of system integration, lack of return of investment in new technology and financial issues during the erection of new smart manufacturing systems and/or during the upgrade of existing industries with smart manufacturing technology. The challenges faced by the smart manufacturing systems and their possible solutions are described below.

11.3.1 Security Issues in Smart Manufacturing

The smart manufacturing system means the use of an integrated network system in a manufacturing system for sharing information between manufacturing or machining units to the end customers. For this purpose, it requires network connectivity and is arranged especially through the internet. Sharing information through the internet requires security of data and information throughout the system in various points with global unique identification and end to end data encryption [12]. And hence, every node of the network should be protected against external attacks and data misuse. The thing that is most importantly considered while designing the networked systems such as smart manufacturing systems is to ensure the security of the system and the overall process [10].

11.3.2 System Integration

Another challenge of the implementation of a smart manufacturing system is the integration of new technology equipment with the existing ones. The compatibility of existing devices to the new devices causes various problems in the implementation of smart manufacturing technologies. The old machinery which is being controlled by some communication protocols might be outdated and new devices may have a different protocol. Also, the machine to machine communication and the interconnectivity of the system requires the better communication system. The recent manufacturing systems require IPv6 connectivity to support more devices connected at the same time.

11.3.3 Interoperability

The interoperability is the ability of different systems to understand and access each other's functions independently. This feature supports the data and information exchange between them independent of the manufacturer of their hardware or software. There are four levels of interoperability of

I4.0, namely operational, systematical, technical and semantic. The operational interoperability correlates CPS with I4.0 and concerns with the conceptual structure of I4.0. Systematical interoperability concerns with standards, guidelines, principle methodologies and models. The technical interoperability distinctly establishes tools and platforms for technical and ICT environment and related software. Semantic interoperability deals with information exchange among different levels of institutions and people [11] .

Without proper matching of the communication protocols and standards, the feature of interoperability may not be achieved efficiently. The dissimilarities between the communication bandwidth, operational frequency, mode of communication, hardware capabilities etc. determine the limitations of interoperability of the system.

11.3.4 Safety in Human–Robot Collaboration

A cobot or lightweight robot is a particular type of robot which can safely and physically interact with humans in the workspace and can work together by introducing new paradigms from Human-Machine interaction (HMI). The international federation of Robotics defines Human-Robot collaboration as the ability of the robot to coordinate with the workers in the industrial environment to perform the specialized tasks. The main considerations should be done with occupational health and safety of personnel working in the site, any hazardous environment should be avoided and necessary occupational health and safety should be maintained. The main attention should be given while implementing the CPS system or industrial robotic systems to minimize any types of mechanical, electrical, thermal, noise, vibration, radiation, material/substance, work environment caused any hazards in the workplace in an industry [12].

11.3.5 Multilingualism

The smart manufacturing systems should be handling multilingual operations which should be able to interpret any instructions given in human language into the machine language to instruct the machine for the desired operation. To make the term smart manufacturing to be realistic and implement AI and advanced technologies in manufacturing system, it should be able to get the instructions directly from the operator either in voice or in text format.

11.3.6 Return of Investment in New Technology

While shifting into another advanced technology in an existing manufacturing system, the financial analysis and return of investment are analyzed very carefully. The additional investment that should be made to adopt newer technology would be compared with the losses in production during an upgrade and the time acquired to recover the return of investment with the income with existing system influences the adaption of newer technology.

11.4 Enabling Mechanisms Toward Smart Manufacturing for Greener Future

11.4.1 Change Manufacturing Processes

11.4.1.1 Change the Nature of Process

Switching from a chemical process to a physical or a biological process often becomes more beneficial for the environment. For example, using a controllable pressure to create the flexible polyurethane foam, rather than a chemical process involving chlorofluorocarbons can reduce GHG emissions [13].

11.4.1.2 Advanced Manufacturing Processes

Intelligent technologies can be used directly to reduce GHG emissions from industrial processes. In the large-scale production of hydrogen, GHG emissions are produced as a side product during the steam reforming and heating of reactors. Electrification technology can reduce GHG emissions and offer flexible and compact heat generation [14].

11.4.1.3 Greener Inputs

Both major and minor ingredients which contribute to the production and are not included in the end product are worth changing. As one of the major sources of producing energy with CO2 as a byproduct, fossil fuels are depleting rapidly. There is an urgent need of finding alternative energy from renewable sources, including hydrogen, biobutanol and bioethanol.

11.4.1.4 Advanced Product Design

Intelligent tools, e.g., Autodesk Inventor and Ansys Granta Selector, have been developed to support the design of green alternatives based on the analysis of material properties, traceability, configuration, and environmental impacts. From the analysis, greener materials and production plans can be optimized.

11.4.2 Reducing Production Waste

In order to reduce waste and emissions, IM technologies can be deployed to improve the efficiency and quality in making existing products, thus controlling the amount of waste to be placed.

11.4.2.1 Lean Manufacturing

Lean manufacturing is a fundamental concept to reduce waste in production with assured quality. Creating continuous material flows from customers to suppliers and using tact time and pull system to manage the workflow provide continuous improvement in all stages of a manufacturing process towards zero-waste [15].

11.4.2.2 Selective Use of Automation

Smart sensors and machines take over parts of what humans used to do, which reduces the dependence of manufacturing processes on active human intervention. Automated process control, human-robot collaboration, and numerically controlled machine tools can effectively reduce human errors and defections, thus reducing wastes.

11.4.2.3 Advanced Manufacturing Process

Additive manufacturing (3D printing) makes a physical object from a three-dimensional digital model. Unlike a subtractive process, where only a small percentage of the material goes into the final part, the material wastes in 3D printing are minimized [16].

11.4.2.4 Greener Cleaning Materials

Cleaning is the source of water waste from manufacturing processes. These impacts can partly be reduced by changing inputs from solvents to water- and bio-based cleaners.

11.4.3 Improving Energy Efficiency

Increased adoption of intelligent technologies that combine advanced manufacturing technologies and intelligent efficiency management is promising to help manufacturers reduce facility-level energy intensity. Besides the aforementioned technologies, including advanced manufacturing processes, greener inputs, advanced product design, lean manufacturing, we identified the following IM technologies as additional enablers for improving energy efficiency.

11.4.3.1 Equipment and Factory Design

Greener high-end equipment has been designed to reduce the energy consumption in cutting and grinding. Moreover, a factory can be continuously improved for better energy efficiency, from the aspect of air conditioning, e.g., arranging air circulation fan and optimizing airflow, lighting, and logistics, e.g., minimizing material transportation distance.

11.4.3.2 Internet of Things (IoT)

The increasing capabilities of internet-connected sensors and the associated IoT techniques provide advanced metering and sub-metering solutions, which substantially improve the performance of energy management systems. Some enterprises have implemented a highly digitalized manufacturing system, which monitors and controls the production and energy usage, projecting a 20% average reduction in energy consumption per industrial facility in 2035.

11.4.3.3 Intelligent Scheduling and Artificial Intelligence

Intelligent scheduling and artificial intelligence help manufacturers unlock energy efficiency from the equipment level to the entire supply chain. For example, the intelligent scheduling process of multiple connected manufacturing systems is modeled for adaptively configuring lines and scheduling the jobs based on the stochastic demands to increase the utility rate of energy [17].

11.4.4 Making Manufacturing Sustainable

Sustainable manufacturing targets to increase operational efficiency, respond to new customers' requirements, and build long-term business viability and success.

11.4.4.1 Remanufacturing

Remanufacturing aims to restore the returned product to full functionality and use it for another lifecycle and has huge potentials for energy and waste reduction. Studies discovered that remanufacturing has the potentials for reducing carbon emissions, improving the product quality, eco-efficiency, and service level.

11.4.4.2 Service-Oriented Manufacturing (SOM)

SOM aims to share manufacturing capabilities and resources through a cloud platform. SOM offers a solution to leverage the intelligent decision-making capacities from the cloud, to provide the most sustainable and robust options in manufacturing. For instance, Plex Manufacturing Cloud is a software that packages the enterprise resource planner (ERP) as a service (SaaS) help manufacturers to organize their manufacturing process more efficiently and intelligently.

11.4.4.3 Reconfigurable Manufacturing

Reconfigurable manufacturing (RM) was initially designed to enhance responsiveness to unpredictable market changes and recently showed its superiority in managing wastewater and emissions using system reconfiguration. Khezri *et al.* proposed an optimization model for sustainable operation in a reconfigurable manufacturing environment with mathematical modeling of hazardous waste and emissions. Meanwhile, combining the IoT technologies and artificial intelligence, machine failure can also be detected and inferred earlier in RMS to reduce the defect rate and to improve manufacturing efficiency [18].

11.5 Supporting Tool/Methods Toward Smart Manufacturing for Greener Future

11.5.1 Life Cycle Assessment and Carbon Pricing Strategies

Life cycle assessment (LCA) is a standardized approach to evaluating the environmental impacts of a product, process or activity by tracking material and energy usage and waste outputs at every life stage. It is critical to have an accounting of inputs, wastes, and their associated costs at each point in the manufacturing process. The actual efficiency greatly depends on the automation level of this process. The assessment can be enhanced by artificial intelligence to pinpoint costs, particularly track costs back to specific manufacturing processes to project future costs. Advanced ICTs enable the real-time monitoring of CO_2 from fuel combustion, and air-flow sensing in various parts of the furnaces. Networks of energy and CO_2 emission can be designed using IoT to enhance services in the manufacturing process and to address the complex energy consumptions [19].

Smart manufacturing can leverage carbon emission as an ingredient to design the entire supply chain. In a consumer viewpoint, producers are not directly motivated to reduce emissions, while consumers, instead, should assume responsibility for choosing the best strategies and policies by showing a preference for eco-friendly producers. IoT techniques and digital supply chain connect the entire supply chain, from raw material suppliers to customers and pricing the products considering the environmental costs to morph the market dynamically. Other pricing strategies to encourage greener development are carbon taxes and emissions trading systems (ETS).

A carbon tax directly sets a price on carbon by defining a tax rate on CO_2 emissions. ETS caps the total level of greenhouse gas emissions and allows those industries with low emissions to sell their extra allowances to larger emitters. Compared to carbon tax, ETS of a carbon tax is not often pre-defined but the carbon price negotiated among enterprises. This enterprise-level carbon market helps to reach the required emission reduction target by keeping the emitters reweighing the investment costs of greener technologies and carbon purchase. More than 40 countries and 20 cities, states and provinces already use carbon pricing mechanisms, with more planning to implement them in the future.

11.5.2 Carbon Capture and Storage

All the aforementioned methods aim to reduce carbon emission, either in a short or long term. Carbon capture and storage (CCS) is identified as an effective technique for compensating the existing CO_2 emissions. Norway was the first nation to take CCS seriously for the implementation of CCS devices on all fossil-fired power plants and declared CCS would be Norway's "Moon mission". Recent developments in polymeric membranes have successfully achieved carbon capture on a demonstration scale [20].

CCS has not yet been widely deployed on a commercial scale owing to a variety of technical, economic and political challenges, which makes the prospects for CCS technologies are at a complete standstill in many countries. Two main cost drivers to CCS: 1) energy penalty to separate and compress the CO2 (costs due to the loss of power generation efficiency), and 2) capital costs (costs required to build, install and operate the equipment). There are several benefits gained from intelligent technologies to address the challenges in CCS: (1) Use digital twins to evaluate electrical load and energy consumptions of CCS technologies. (2) Use the IoT to monitor and control plant contribution to CO2 emissions. Vatopoulos and Tzimas have designed a mathematical model as a digital twin to evaluate three CCS technologies, in terms of, oxy-combustion, amine scrubbing, and chemical looping in a cement-making process. In comparison, the chemical looping shows a lower energy consumption than amine scrubbing with 18% increased efficiency.

11.6 Conclusions

Smart manufacturing has emerged as a transformative force in the pursuit of a greener and more sustainable future. This paper has provided insights into the concept of smart manufacturing and its potential to drive environmental sustainability in the manufacturing sector.

One of the key benefits of smart manufacturing is its ability to optimize resource efficiency. By utilizing advanced technologies such as Internet of Things (IoT), artificial intelligence (AI), and data analytics, manufacturers can monitor and analyze real-time data from production processes. This enables them to identify inefficiencies, optimize energy consumption, and reduce waste generation. By implementing predictive maintenance and condition monitoring techniques, companies can also minimize downtime and prolong the lifespan of equipment, thereby reducing the need for replacements and conserving resources.

Furthermore, smart manufacturing facilitates the integration of renewable energy sources. By leveraging IoT and AI technologies, manufacturers can dynamically manage energy consumption and production, optimizing the use of renewable energy. This integration not only reduces reliance on fossil fuels but also contributes to the reduction of greenhouse gas emissions.

Additionally, smart manufacturing systems enable the implementation of sustainable practices throughout the product lifecycle. From design and production to distribution and end-of-life disposal, smart manufacturing technologies provide opportunities to incorporate eco-friendly materials, improve recyclability, and reduce environmental impact. Real-time tracking and traceability systems can also enhance supply chain transparency, ensuring compliance with environmental regulations and ethical sourcing practices.

In conclusion, smart manufacturing offers immense potential for driving a greener future in the manufacturing industry. By harnessing the power of digital technologies, automation, and data analytics, manufacturers can optimize resource efficiency, reduce waste, integrate renewable energy sources, and implement sustainable practices. Embracing smart manufacturing not only enhances environmental sustainability but also brings about cost savings, improved operational efficiency, and enhanced competitiveness in the market. As the world continues to prioritize sustainability, the adoption of smart manufacturing is a crucial step toward a greener and more sustainable future.

References

1. Nordhaus, W., Integrated assessment models of climate change. *NBER Rep.*, 3, 16–20, 2017.
2. Geissdoerfer, M., Savaget, P., Bocken, N.M., Hultink, E.J., The Circular Economy–A new sustainability paradigm? *J. Clean. Prod.*, 143, 757–768, 2017.
3. Sendler, U., *The internet of things: Industrie 4.0 unleashed*, Springer, 2017.
4. Zhou, J., Zhou, Y., Wang, B., Zang, J., Human–cyber–physical systems (HCPSs) in the context of new-generation intelligent manufacturing. *Engineering*, 5, 4, 624–636, 2019.
5. Fraga-Lamas, P., FernáNdez-CaraméS, T.M., Blanco-Novoa, O., Vilar-Montesinos, M.A., A Review on Industrial Augmented Reality Systems for the Industry 4.0 Shipyard. *IEEE Access*, 6, 13358–13375, 2018, doi: 10.1109/ACCESS.2018.2808326.
6. Lee, E.A., The Past, Present and Future of Cyber-Physical Systems: A Focus on Models. *Sensors*, 15, 3, 4837–4896, Mar. 2015, doi: 10.3390/s150304837.
7. Ahuett-Garza, H. and Kurfess, T., A brief discussion on the trends of habilitating technologies for Industry 4.0 and Smart manufacturing. *Manuf. Lett.*, 15, 60–63, Jan. 2018, doi: 10.1016/j.mfglet.2018.02.011.
8. Mabkhot, M.M., Al-Ahmari, A.M., Salah, B., Alkhalefah, H., Requirements of the Smart Factory System: A Survey and Perspective. *Machines*, 6, 2, 23, Jun. 2018, doi: 10.3390/machines6020023.
9. ElMaraghy, H.A., Flexible and reconfigurable manufacturing systems paradigms. *Int. J. Flex Manuf. Syst.*, 17, 4, 261–276, Oct. 2005, doi: 10.1007/s10696-006-9028-7.
10. Thoben, K.-D., Wiesner, S., Wuest, T., 'Industrie 4.0' and Smart Manufacturing – A Review of Research Issues and Application Examples. *Int. J. Autom. Technol.*, 11, 1, 4–16, Jan. 2017, doi: 10.20965/ijat.2017.p0004.

11. Lu, Y., Kaushal, N., Denier, N., Wang, J.S.-H., Health of newly arrived immigrants in Canada and the United States: Differential selection on health. *Health Place*, 48, 1–10, Nov. 2017, doi: 10.1016/j.healthplace.2017.08.011.
12. Gualtieri, L., Palomba, I., Wehrle, E.J., Vidoni, R., The opportunities and challenges of SME manufacturing automation: safety and ergonomics in human–robot collaboration. *Ind. 4.0 SMEs Challenges Opportunities Requir.*, 105–144, 2020.
13. Yang, J., Ye, Y., Li, X., Lü, X., Chen, R., Flexible, conductive, and highly pressure-sensitive graphene-polyimide foam for pressure sensor application. *Compos. Sci. Technol.*, 164, 187–194, 2018.
14. Wismann, S.T., *et al.*, Electrified methane reforming: A compact approach to greener industrial hydrogen production. *Science*, 364, 6442, 756–759, 2019.
15. Leong, W.D., Lam, H.L., Ng, W.P.Q., Lim, C.H., Tan, C.P., Ponnambalam, S.G., Lean and green manufacturing—a review on its applications and impacts. *Process Integr. Optim. Sustain.*, 3, 5–23, 2019.
16. Campbell, T., Could 3D printing change the world?: Technologies, potential, and implications of additive manufacturing, 2011.
17. Li, X., Bayrak, A.E., Epureanu, B., II, Koren, Y., Real-time teaming of multiple reconfigurable manufacturing systems. *CIRP Ann.*, 67, 1, 437–440, 2018.
18. Epureanu, B., II, Li, X., Nassehi, A., Koren, Y., Self-repair of smart manufacturing systems by deep reinforcement learning. *CIRP Ann.*, 69, 1, 421–424, 2020.
19. Liu, W., Peng, T., Tang, R., Umeda, Y., Hu, L., An Internet of Things-enabled model-based approach to improving the energy efficiency of aluminum die casting processes. *Energy*, 202, 117716, 2020.
20. Han, Y. and Ho, W.W., Recent advances in polymeric membranes for CO_2 capture. *Chin. J. Chem. Eng.*, 26, 11, 2238–2254, 2018.

Index

Printed in the USA/Agawam, MA
September 16, 2026

910661.072